Lecture Notes in Artificial Intelligence

Subseries of Lecture Notes in Computer Science
Edited by J. G. Carbonell and J. Siekmann



Lecture Notes in Computer Science
Edited by G. Goos, J. Hartmanis, and J. van Leeuwen

Springer
*Berlin*
*Heidelberg*
*New York*
*Hong Kong*
*London*
*Milan*
*Paris*
*Tokyo*

Michael Beetz

# Plan-Based Control of Robotic Agents

Improving the Capabilities of Autonomous Robots

Springer

Series Editors

Jaime G. Carbonell, Carnegie Mellon University, Pittsburgh, PA, USA
Jörg Siekmann, University of Saarland, Saarbrücken, Germany

Author

Michael Beetz
Technische Universität München
Institut für Informatik IX
Boltzmannstr. 3, 85748 Garching b. München
E-mail: beetzm@in.tum.de

Cataloging-in-Publication Data applied for

A catalog record for this book is available from the Library of Congress

Bibliographic information published by Die Deutsche Bibliothek
Die Deutsche Bibliothek lists this publication in the Deutsche Nationalbibliographie; detailed bibliographic data is available in the Internet at <http://dnd.ddb.de>.



CR Subject Classification (1998): I.2, C.2.4, C.3

ISSN 0302-9743
ISBN 3-540-00335-5 Springer-Verlag Berlin Heidelberg New York



Springer-Verlag Berlin Heidelberg New York,
a member of BertelsmannSpringer Science+Business Media GmbH

http://www.springer.de


Printed in Germany

Typesetting: Camera-ready by author, data conversion by Christian Grosche, Hamburg
Printed on acid-free paper SPIN: 10871746 06/3142 5 4 3 2 1 0

*For Annette,*
*Nicola, and Fabian*

# Preface

Robotic agents, such as autonomous office couriers or robot tourguides, must be both reliable and efficient. This requires them to flexibly interleave their tasks, exploit opportunities, quickly plan their course of action, and, if necessary, revise their intended activities. In this book, we describe how ***structured reactive controllers*** (SRCs) satisfy these requirements. The novel feature of SRCs is that they employ and reason about plans that specify and synchronize *concurrent percept-driven* behavior. Powerful control abstractions enable SRCs to integrate physical action, perception, planning, and communication in a uniform framework and to apply fast but imperfect computational methods without sacrificing reliability and flexibility. Concurrent plans are represented in a transparent and modular form so that automatic plan management processes can reason about the plans and revise them.

The book makes three main contributions. First, it presents a plan representation that is capable of specifying flexible and reliable behavior. At the same time, the plan representation supports fast and robust execution time plan management. Second, it develops *Probabilistic Hybrid Action Models* (PHAMs), a realistic causal model for predicting the behavior generated by modern concurrent percept-driven robot plans. PHAMs represent aspects of robot behavior that cannot be represented by most action models used in AI planning: the temporal structure of continuous control processes, their non-deterministic effects, and several modes of their interferences. Third, it describes XFRMLEARN, a system that learns structured symbolic navigation plans. Given a navigation task, XFRMLEARN learns to structure continuous navigation behavior and represents the learned structure as compact and transparent plans. The resulting plans support action planning and opportunistic task execution.

We present experiments in which SRCs are used to control two autonomous mobile robots. In one of them an SRC controlled the course of action of a museum tourguide robot that has operated for 13 days, more than 94 hours, and has performed about 3200 execution time plan management operations.

In working on this book, I have greatly benefited from the help and support of many people. Here are those I would like to thank especially.

First, I would like to thank Prof. Armin B. Cremers and Prof. Bernd Radig for their support and for providing such excellent research environments. I would also like to express my gratitude to Prof. Armin B. Cremers, Prof. Malik Ghallab, Prof. Rainer Manthey, Prof. Martha Pollack, and Prof. Bernd Radig for being the readers of my Habilitation thesis. Joachim Hertzberg has been a perfect (part-time) office mate, friend, and discussion partner.

Carrying out successful research in autonomous robot control is not possible without being a member of a team. I had the luck of joining one of the best teams: the RHINO team. I would like to thank especially the following members and alumni of the RHINO team: Tom Arbuckle, Thorsten Belker, Maren Bennewitz, Wolfram Burgard, Dieter Fox, Henrik Grosskreutz, Dirk Hähnel, Dirk Schulz, Gerhard Lakemeyer, and Sebastian Thrun.

After leaving the RHINO group I joined the Intelligent Autonomous Systems Group at the Technische Universität München, another great robotics team. Here, I would especially like to thank my colleagues Sebastian Buck, Robert Hanek, Thorsten Schmitt, Derik Schröter, and Freek Stulp.

Special thanks go to those I have collaborated with while they carried out their Master's thesis research, most notably Thorsten Belker, Maren Bennewitz, and Henrik Grosskreutz, but also Arno Bücken, Markus Giesenschlag, Markus Mann, Hanno Peters, and Boris Trouvain. I had also the pleasure to work with Wolfram Burgard, Dieter Fox, and Dirk Schulz on the integration of probabilistic state estimation into high-level control, with Tom Arbuckle on the integration of image processing, and with the Minerva team on the 1998 Museum tourguide project.

Most importantly, I want to thank my wife Annette and children Nicola and Fabian for their love and support and putting up with me writing this manuscript.

October 2002 Michael Beetz

# Contents

# 1. Introduction

Over the last years, the perception, navigation, planning, and reasoning capabilities of autonomous robots have been considerably improved. As a consequence, autonomous service robots acting in human working environments such as autonomous office couriers (Simmons et al., 1997b; Beetz, 2001) or museum tour-guides (Burgard et al., 1998; Thrun et al., 2000) have become challenging test beds for developing and testing computational models of competent agency.

Controlling autonomous robots entails specifying how the robots are to react to perceptual input in order to accomplish their goals (McDermott, 1992a). AI-based robotics researches specializations of this control problem. One of these specializations' key challenges is the creation of autonomous robots that perform everyday activity in real-world environments over extended periods of time. Everyday activity requires such robots to accomplish complex routine jobs in changing environments where they face familiar but often not completely known situations. To ensure continuing utility, the robots must manage their tasks and execute them flexibly and reliably.

We believe that in order to generate such flexible and goal-directed activity, autonomous robots need to exhibit activity patterns that are typical for human everyday activity. Such activity patterns include the flexible interleaving of different jobs and the exploitation of unexpected opportunities. Exhibiting such patterns also requires the robots to quickly plan their course of action and revise their intended activities in order to resolve conflicts between interfering jobs. To perform their everyday activity reliably, robots even have to reconsider their course of action in the light of new information.

Our overall research goal is the investigation of computational models for accomplishing everyday activity and their realization on autonomous robots. We take, as a prototypical example, a robot office courier. Essentially, a robot office courier satisfies user requests of the form "get an item at location *s* and bring it to location *d*." As a consequence, the robot's task management mechanism must decide the order, in which objects are to be picked up and delivered. At the same time, the robot has to decide when batteries should be reloaded and when to re-localize itself in order to avoid getting lost and to avoid delivering items at the wrong places. The decision making for the task management also must ensure that the robot meets the deadlines for the user requests and is not carrying items that can be confused. Besides the reliable satisfaction of user requests, the fast completion of the deliveries is another important objective.

Task management is complicated by the fact that the robot may — at any point in time — perceive additional information that might warrant changes in its intended course of action while it performs its activities. Consider the following situation. In the midst of carrying out a planned delivery tour the robot recognizes, while passing an office, that a door that it previously assumed to be closed is in fact open. What should the robot do? Should it simply ignore the open door and continue its intended course of action? Or, should it consider the open door as an opportunity to achieve its jobs in a better way? A robot, passing offices at walking speed, must decide quickly and based on incomplete and uncertain information. Taking an open door as an opportunity might yield subtle side effects in the future part of the robot's intended course of action. For example, taking an opportunity might cause delays or affect the items that the robot will be carrying.

In this book we develop a task management mechanism for autonomous robots that can make such execution time decisions quickly, competently, and with foresight. The mechanism controls a physical robot that is equipped with general and reliable navigation skills, routines for estimating certain states of the environment, basic image processing capabilities, and some basic means for communicating and interacting with the robot.

## 1.1 The Challenges of Controlling Robot Office Couriers

The properties of robotic agents, their jobs, and the environments they are acting in imply several challenges for the task management and task execution mechanisms of robotic office couriers.

First, the robot office courier must satisfy multiple complex user requests that are dynamically added, revised, or canceled. Sometimes a user request cannot be accomplished without taking into account the other activities of the robot because the intended activities for accomplishing each individual user requests might be incompatible. Thus, accomplishing all user requests reliably requires the robot controller to infer possible interactions and to avoid or exploit them.

A physical robot that is to accomplish multiple and complex tasks under dynamically changing circumstances typically yields a broad variety of execution failures including missed deadlines whilst exploiting incidental opportunities and the confusion of items that are to delivered. In order to anticipate and forestall these kinds of failures, the task management mechanism needs very realistic and detailed models of the control routines that perform the tasks and the behavior produced by them.

The second challenge consists of the task management being tightly coupled to the robot's perceptual and action apparatus. The action apparatus includes physical actions, such as navigation, communication and interaction with people, reasoning, and planning. The nature of these different kinds of actions and interactions is very different. For example, physical actions are mostly movements. Movements have continuous effects with temporal and spatial structure. These effects should be continually monitored and exceptional situations must be reacted to. User interactions, on the other hand, have very different characteristics. Such interactions can be initiated by the robot, but after their initiation the interactions are to be continued by other agents, and therefore, interaction is only partially under the control of the robot. Clearly, the flexible and reliable performance of such different actions and interactions requires mechanism-specific control patterns.

The same holds for the different components of the perceptual apparatus. Most autonomous robots today are equipped with passive sensors, such as range sensors. Typically, range sensors continually measure the distance to the closest obstacles. Such sensors and the corresponding sensor data processing routines provide the task management mechanism with an asynchronous stream of percepts, which might require reconsideration of the course of action. Many robots also employ active perception mechanisms, where the control system is requesting a specific piece of information that is to be selectively acquired. A large fraction of image processing tasks are of this sort. To get reliable results and process images at a satisfactory speed, the control system should avoid the request of general information, such as "what is in the camera view?" and rather ask specific questions in specific contexts, such as "is the door I am looking at open?". Again the coupling of the plan management system with passive perception routines is very different from the one with active routines.

We conjecture that plan management and execution cannot view the intended courses of actions as partially ordered sets of atomic actions fully under the control of the agent without losing important means for the control and coordination of the robot's different mechanisms. Plans for the flexible and reliable satisfaction of user requests must employ more powerful and feedback driven control patterns and allow for a tight coupling between sensing actions and physical actions — even at a high level of abstraction.

The third challenge is concerned with the integration of plan management and execution. A competent robotic agent must monitor the execution of its intended course of action and change the action course, if necessary. This requires the robot to continually decide when to revise which parts of its intended activities. In addition, reasoning more carefully about the consequences of these decisions at the cost of arriving at control decisions later reduces the risk of making suboptimal plan management decisions. This also implies that the agent has to decide how long it intends to reason before committing to a decision. Performing execution time plan management also requires the smooth installment of plan revisions and the successful continuation of the intended activity.

Finally, the fourth challenge is that a competent robotic agent must be capable of effectively managing and executing its intended course of action based on uncertain information. Typically, the robot is uncertain about the accurate state of the environment, about the durations and outcomes of its intended activities, and the occurrence of exogenous events. This uncertainty has two kinds of consequences for the robot courier. First, it should use plans that specify flexible and reliable behavior. These flexible plans, however, violate assumptions underlying most action models used for planning — in particular, the assumptions about the irrelevance of the temporal structure of actions and the exclusion of interferences between concurrent actions. Second, for managing the tasks it should take the various kinds of uncertainty into account and reason through contingencies whose likelihoods are a priori unknown and varying in order to improve the expected performance of its activity. These kinds of uncertainty include uncertainty about the accurate state of the environment, the durations and outcomes of actions, and the occurrence of exogenous events.

There are several other important challenges that must be met by successful realizations of flexible and reliable robot office couriers. These challenges include among others object recognition and object manipulation. We will not address these issues in this book because our robot does not provide the necessary perceptual and manipulation repertoire.

## 1.2 The Control Problem

In general, the control problem for autonomous robot is the generation of effective and goal-directed control signals for the robot's perceptual and effector apparatus within a feedback loop. Within each cycle the robot has only a fixed amount of computational resources available for selecting the control signals for the next cycle.

In this book we investigate a specialization of this control problem, in which the robot generates the control signals by maintaining and executing a plan that is effective and has a high expected utility with respect to the robot's dynamically changing belief state. In this control paradigm plan management operations and plan execution are performed in concurrent threads of control. The plan management operations precompute control decisions and install them into the plan. The plans generate the control signals for the subsequent feedback cycles.

Plan-based control is often more economical than the computation of the next control signals from scratch. This happens in particular if the robot's tasks are complex and possibly interfering and the environment is changing and only partly accessible using the robot's sensors. Under these conditions, the belief state of the robot may change from cycle to cycle and the choice of the control signals with the highest expected utility might require to reason through all possible future consequences of the decision. The use of plans helps to mitigate this situation in at least two ways. First, it decouples computationally intensive control decisions from the time pressure that dominates the feedback loops. Precomputed control decisions are reconsidered only if the conditions that have justified the decisions have changed. Second, plans can be used to focus the search for appropriate control decisions. They can neglect control decisions that are incompatible with its intended plan of action.

While plan management aims at reducing the number of computationally expensive reasoning tasks that have to be performed, the reasoning tasks themselves include robot planning, the generation, refinement, and revision of robot plans based on prediction (McDermott, 1992a). Unfortunately, robot planning is a computationally very expensive reasoning task.

In Computer Science it is common to characterize the computational problems a program can solve through the language in which the input for the program is specified. For example, we distinguish compilers for regular and context-free programming languages. The same is true for plan-based control of agents. Typically, planning problems are described in terms of an initial state description, a description of the actions available for the agents, their applicability conditions and effects, and a description of the goal state. The planning system is to compute a sequence of action that transforms any state satisfying the initial state description in another state that satisfies the given goals.

The three components of planning problems are typically expressed in some formal language. The problem solving power of the planning systems is characterized by the expressiveness of the languages for the three inputs. Some classes of planning problems are entirely formulated in propositional logic while others are formulated in first order logic. We further classify the planning problems with respect to the expressiveness of the action representations that they use; whether they allow for disjunctive preconditions, conditional effects, quantified effects, and model resource consumption. Some planning systems even can solve planning problems that involve different kinds of uncertainty.

The crux with planning problems being formulated this way is, as McDermott (1992a) points out, that those formulations result in planning problems that are both too hard and too easy. They are too hard because even severely simplified versions are unsolvable or — at the very best — computationally intractable. They are too easy because the resulting action sequences are insufficient, sometimes even deceptive, as behavior specifications for autonomous robots.

On the other hand, the fact that people manage and execute their daily tasks effectively suggests, in our view, that the nature of everyday activity should permit robotic agents to make assumptions that simplify the computational problems that need to be solved to produce competent activity. As

Horswill (1996) puts it, everyday life must provide us with some loopholes, structures and constraints that make everyday activity tractable.

We believe that in many applications of robotic agents that are to operate in human working environments the following assumptions are valid and constitute loopholes that allow us to simplify the computational problems that are part of controlling a robot:

1. Robotic agents are familiar with the activities for satisfying single user requests and the situations that typically occur while carrying out these activities. They carry out everyday activities over and over again and are confronted with the same kinds of situations many times. As a consequence, conducting individual everyday activities is simple in the sense that they do not require a lot of plan generation from first principles.
2. Appropriate plans for satisfying multiple user requests can be determined in a greedy manner. The robot can first determine a default plan for accomplishing multiple requests through simple and fast heuristic plan combination methods. The robot can avoid the remaining interfering interactions between its sub-activities by predicting and forestalling these interactions.
3. Robotic agents can monitor the execution of their activities and thereby detect situations in which their intended course of action might fail to produce the desired effects. If such situations are detected, the robots can adapt their intended course of action to the specific situations they encounter.

The working hypothesis underlying this research is that robotic agents can competently accomplish complex and dynamically changing everyday tasks in human working environments if they use general, flexible *routines*, *recognize* problems, opportunities, and interactions between sub-activities, and *adapt* their activity if necessary.

Designing a control system according this hypothesis implies that the plan management operations have to reason through highly conditional and flexible plans and perform execution time plan revisions. As a consequence, researchers have shied away from implementing robot control systems that fully reflect the hypothesis because of the seeming computational complexity of the reasoning tasks.

The nature of the assumptions that we make also deserves some further discussion because it is very different from the nature of those made in classical planning. In classical planning assumptions are made about the robot being omniscient, being the only agent acting in the environment, having full control about its actions, and carrying out one action at a time. We have learned that even making such strong assumptions does not prevent the classical planning problem from being computationally intractable (Bylander, 1991). Yet, the resulting plans are brittle and most of the time inappropriate to specify competent robot behavior.

Our assumptions, on the other hand, are about the properties of jobs, the plan schemata available to the agent, and about the frequency and nature of the interactions between sub-plans. Consider the job of a robot office courier. Performing individual deliveries is routine. The robot goes to the place where the item is to be picked up, retrieves the item, goes to the destination to unload it. There are many different kinds of situations that a robot might be confronted with. Doors might be closed, people might not be at their desks, people might not be responding, and so on. These situations are typical for individual deliveries and agents have been confronted with these situations often before.

Because of their frequency, a robot can learn from experience how to deal with such situations appropriately and precompile the learned decisions as situation-specific default decisions into the plans. Also, when having to satisfy a set of requests, the interactions between the requests are typically weaker than those in block stacking problems. The individual delivery requests can be satisfied in many orders. The robot can order them to obtain shorter delivery tours, rearrange them to satisfy more urgent requests earlier, postpone deliveries that can at that time not be completed, insert opportunistic steps, and so on. The point is: because these interactions are weak, the robot can plan a first default delivery tour quickly using heuristic methods and improve the tour after having started carrying it out. Also, the robot can replan its tour in dynamically changing situations.

As we will see in the remainder of this book, competent plan management can be performed concurrently with the execution of scheduled activity without slowing down the execution of delivery plans.

## 1.3 The Computational Model

If we accept the assumptions about the nature of activity that is carried out by autonomous robots in human working environments as reasonable, then a high-level controller that exploits these concepts can be based upon the following computational principles:

1. The controller has a library of routine plan schemata for everyday activities. Primary activities are concurrent, percept-driven, process-oriented, flexible, robust, general. Primary activities are appropriate only under certain assumptions.
2. The controller has fast heuristic methods for putting plans together from routine activities. It is able to project problems that are likely to occur in its intended course of action and revise its course of action to avoid these problems.
3. The controller performs execution time plan management. It runs processes that monitor the beliefs of the robot and are triggered by certain

belief changes. These processes revise the primary activities while the primary activities are carried out.

Our control model distinguishes between *primary* and *constraining (secondary)* sub-plans, a distinction first introduced by McDermott (1977).[1] A primary plan is one whose intent may be specified in isolation, for example, achieve a certain state of affairs. Secondary plans are executed successfully by executing some other task in a particular way. The execution of secondary plans amounts to monitoring and if necessary, influencing the execution of primary tasks. An example of a secondary task constraining the execution of a primary task is carrying out delivery tours without letting the battery voltage getting too low.

The computational principles are realized through nested *self-adapting plans* that contain so-called *plan adaptors.* Plan adaptors are triggered by specific *belief changes.* Upon being triggered the adaptors decide whether adaptations are necessary and if so perform them. Let us consider the following two self-adapting plans as examples:

**With plan adaptor Whenever** the robot receives new requests
**if** its set of Jobs $J$ **might** interfere
**then** it **changes the course of action**
to avoid these interferences
**Concurrent reactive plan**

**With plan adaptor Whenever** the robot detects an open door
that was assumed to be closed
**if** this situation **is an opportunity**
**then** it **changes its course of action**
to make use of the opportunity
**Concurrent reactive plan**

There are two reasons why we want to represent plan management mechanisms explicitly as part of the plan. The first one is expressiveness. Plan adaptors enable the robot to specify revisions of behavior specifications in addition to specifying behavioral responses. Besides local criteria for choosing the action plan, adaptors can take global criteria, such as interactions between plan steps, into consideration. It is difficult to accomplish such functionality without plan management capability. The second reason for representing plan management mechanisms explicitly rather than hiding them

[1] Secondary plans have been called policies. We will try not to use the term policy to avoid confusion with the term policy as it is used in the literature on solving Markov decision problems. Instead we will use the term constraining plan or constraining task.

in separate modules is that this way the robot can make the interactions between plan management and execution explicit. In addition to the interaction the decision logic underlying the plan adaptors can also be made explicit. Explicit representation of the interactions between plan management and execution and the plan revision logic are the representational prerequisites for reasoning about execution time plan adaption and automatically learning plan adaption mechanisms.

The execution of such self-adapting plans confronts the robot with a series of challenges that the robot's plan management mechanisms have to meet. First, the plan adaptors must reason about and manipulate plans that specify concurrent reactive behavior. Second, in order to decide whether or not a plan adaptation should be performed the robot must accurately predict how the world might change as the robot executes its plan. This requires accurate probabilistic prediction of the continuous physical effects of executing concurrent reactive plans based on uncertain and incomplete information. Third, because the applicability of plan adaptations is tested while the plan gets executed, the robot must decide quickly. Fourth, because plan execution proceeds while the plan adaptor reasons about candidate revisions of the plan, the resulting revisions must be integrated smoothly into the ongoing activity.

We will describe and discuss how these challenges can be met in the body of this book.

The self-adapting plans are realized as *structured reactive controllers* (SRCs) (Beetz, 1999). SRCs are collections of concurrent reactive control routines that adapt themselves to changing circumstances during their execution by means of planning. SRCs employ and reason about plans that specify and synchronize *concurrent percept-driven* behavior. The plans are represented in a transparent and modular form so that automatic planning techniques can reason about them and revise them. SRCs are implemented in RPL (Reactive Plan Language) (McDermott, 1991). The reactive control structures that are provided by RPL are also used to specify how the robot is to react to changes of its beliefs.

## 1.4 An Adaptive Robotic Office Courier

Let us now consider how a robotic office courier working according these control principles will manage and execute its delivery tours. While carrying out the delivery tour, the robot reschedules its delivery tasks; and symbolically predicts the course of action, and forestalls execution failures. In the course of action the robot exhibits several behavior patterns that are typical for everyday activity.

In this example, we will use the mobile robot RHINO (see Figure 1.1), a B21 robot manufactured by "Real World Interface Inc." (RWI). RHINO is equipped with on-board PCs connected to a computer network via a tetherless radio link. The sensor system consists of 24 ultra-sonic proximity sensors, 56 infrared sensors, 24 tactile sensors, two laser range finders, and a stereo color camera system mounted on a pan-tilt unit. The range sensors measure the distance to the closest obstacles. These distance measures are then interpreted to detect obstacles, to build maps of the environment, to estimate the robot's position within the map, and to estimate the opening angles of doors.

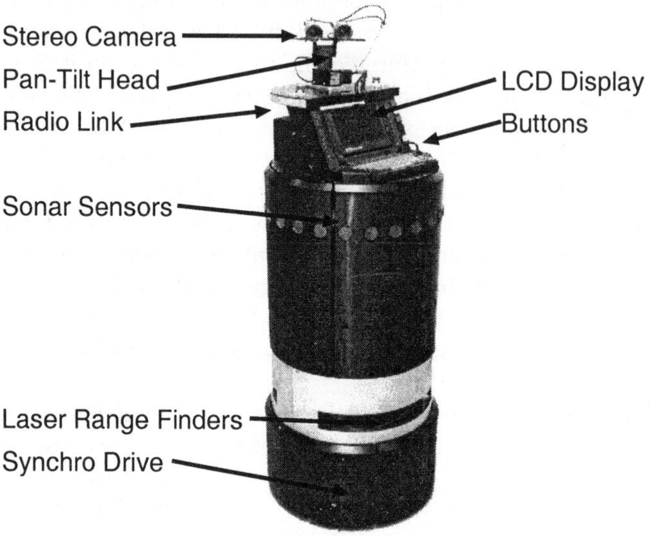


**Fig. 1.1.** The autonomous mobile robot RHINO with its main sensors and actuators

The stereo color cameras enable RHINO to recognize objects in its environments. The radio link is the robot's means for transmitting and receiving digital information. Among other things, the radio link enables the robot to send and receive "electronic mail" in order to communicate with people in its environment.

RHINO is equipped with a library of ***routine plans*** for its standard tasks including delivering items, picking up items, and navigating from one place to another one. The routine plan for deliveries specifies that RHINO is to navigate to the pickup place, wait until the letter is loaded, navigate to the destination of the delivery, and wait for the letter to be unloaded. The routine plans also contain safe and fast navigation routines that dynamically adapt the travel mode to the surroundings that the robot is navigating through (Beetz et al., 1998). Using adaptive navigation plans the robot can drive slowly in offices (because they are cluttered) and quickly in the hallway because they are typically only sparsely cluttered.

Consider the following delivery tour that is carried out by RHINO using its plan-based controller. RHINO receives two requests (see Figure 1.2): "put the red letter on the meeting table in room A-111 on the desk in A-120." and "deliver the green book from the librarian's desk in room A-110 to the desk in A-113."

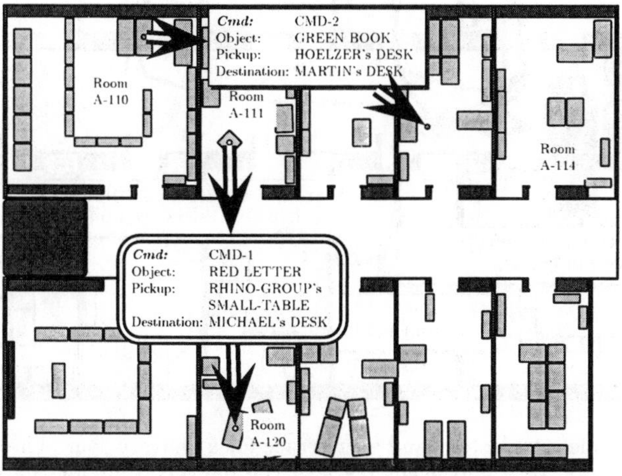


**Fig. 1.2.** The figure shows RHINO's operating environment for the robot office courier application. The environment consists of a small hallway with adjacent offices, a library, and a classroom. The figure also depicts two delivery requests

Whenever RHINO carries out a set of delivery jobs, it quickly computes an default schedule (Beetz, Bennewitz, and Grosskreutz, 1999). RHINO's initial schedule is to deliver the red letter first and the green book afterwards. Proposing a schedule implies making assumptions about whether doors are open and closed. In our example run RHINO assumes that the rooms A-110, A-111, A-113, and A-120 are open. To adapt its schedule flexibly, RHINO monitors the scheduling assumptions while performing its deliveries: whenever it passes a door it estimates the opening angle of that door and revises the schedule if necessary.

Figure 1.3, 1.4(a), and 1.4(b) picture three qualitatively different event traces that the SRC has produced while carrying out the delivery jobs under different circumstances. We will explain the one depicted in Figure 1.3 in more detail. Initially, all doors in the environment are open. RHINO starts with the delivery of the red letter and heads to the meeting table in A-111 where the letter is loaded **(step 1)**. Then the door of A-120 is closed **(step 2)**. Thus, when RHINO enters the hallway to deliver the red letter at Michael's desk it estimates the opening angle of the door of room A-120 and detects that the door has been closed **(step 3)**. A failure is signaled.

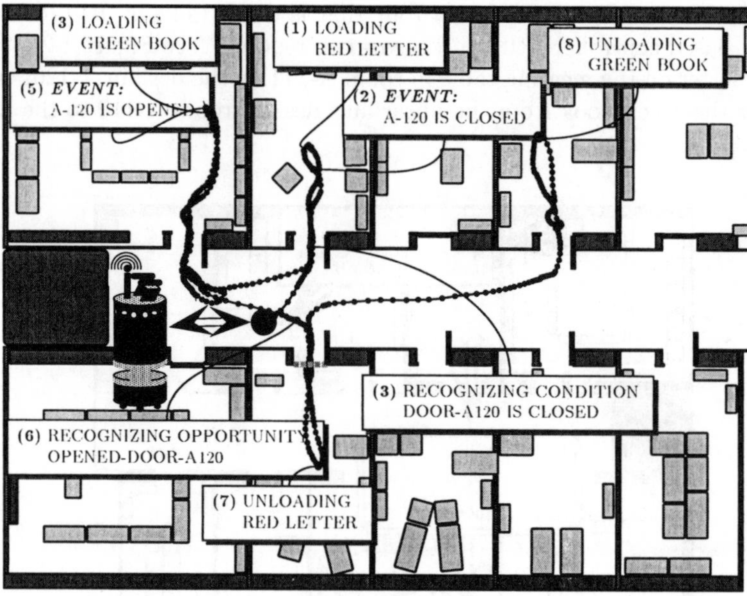


**Fig. 1.3.** Complete trajectory and event trace for a delivery tour. The figure shows the complete trajectory and event trace of the delivery tour for the two delivery requests from figure 1.2. In this event trace office A-120 is closed while the robot is in office A-111 and opened again while it is in office A-110

Since RHINO does not know when room A-120 will be open again, it revises the schedule such that it now delivers the green book first and accomplishes the failed delivery as an opportunity. Thus, RHINO navigates to the librarian's desk in A-110 to pick up the green book to room A-113 **(step 4)**. At this moment room A-120 is opened again **(step 5)**.

As RHINO heads towards A-113 to deliver the green book it passes room A-120 **(step 6)**. At this point the door estimation process signals an opportunity: A-120 is open! To cope with the situation that might constitute an opportunity, RHINO does two things. First, it makes a quick decision to interrupt its current delivery to complete the delivery of the red letter because A-120 is often closed. Second, it starts at the same time reasoning through the future consequences of taking the opportunity, a computational process that might take a considerable amount of time. If RHINO's prediction implies a substantial chance of taking the opportunity causing execution failures later, then it revises its plan again to ignore the opportunity and continue with its original plan. In the other case, if no problems are predicted it completes the opportunity first and then continues with the remaining parts of its plan. In our scenario no problem is predicted and the robot keeps taking the opportunity. After the delivery of the red letter is completed **(step 7)**, RHINO continues the delivery of the green book **(step 8)**.

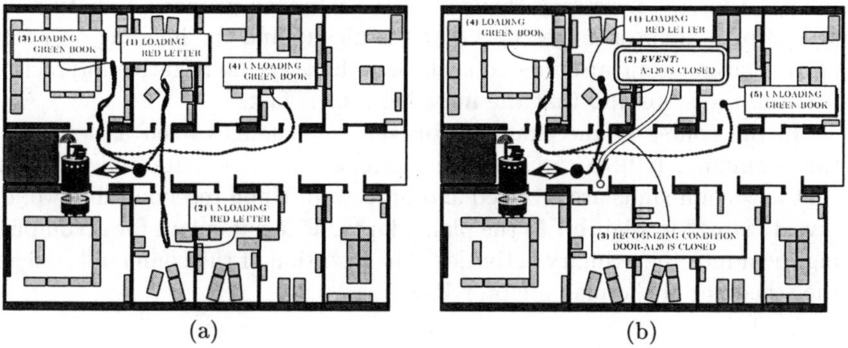


**Fig. 1.4.** Two event traces for the same delivery tour in different situations. Subfigure (a) shows an event trace for the two deliveries if all doors are open. Subfigure (b) shows an event trace for the case that office office A-120 is closed while the robot is in A-111 and stays closed afterwards

Figures 1.4(a) and 1.4(b) show the behavior generated by the same SRC if all doors stay open (Figure 1.4(a)) and if A-120 is closed but not opened again (Figure 1.4(b)).

with plan adaptors

(1) *integrate command revisions* **(P-1)**

(2) *replan/reschedule when necessary* **(P-2)**

(3) *reschedule when A-110, A-111, A-113, or A-120 are closed* **(P-3)**

primary activities

CMD-1: *Deliver the red letter ...*

CMD-2: *Deliver the green book ...*

*:order* CMD-1 CMD-2

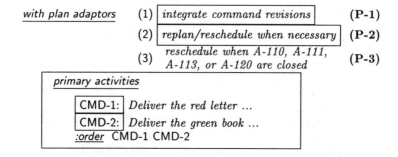

**Fig. 1.5.** The figure shows the top-level structure of the SRC after the plans for the two delivery requests have been inserted. The SRC also contains a plan adaptor that triggers a rescheduling process when one of the doors that have to be passed on the delivery tour is detected to be closed

The self-adapting plan that accomplishes these behaviors consists of the *primary activities* that specify the course of action for satisfying the user requests — the delivery tour. The constraining plans comprise two stabilizing plans: one for reloading the battery when it gets empty and one for re-localizing the robot when it gets lost. They also contain two general plan adaptors. The first one makes sure that the intended course of action contains delivery tasks for every actual user request and no unnecessary delivery tasks. Thus, whenever a user request is added, deleted, or revised, the plan adaptor performs the necessary changes on the intended primary activities. The second plan adaptor makes sure that the pickup and the handing over steps of the delivery tour are ordered such that there are no negative interactions between the steps and that the intended tour is fast.

In the course of the example run the SRC changes itself as follows. In the beginning, RHINO carries out no primary activities. Its policies ensure that new commands are received and processed. Upon receiving the two jobs (event "start" in Figure 1.3) the plan adaptor **P-1** puts plans for accomplishing them into the primary activities. The insertion of the commands triggers the scheduler of the plan adaptor **P-2**.

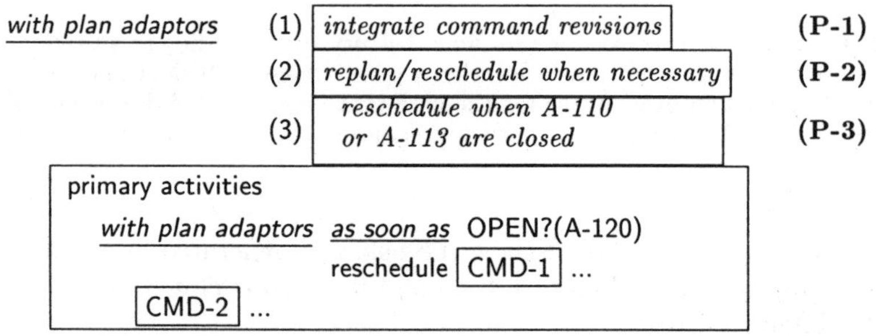


**Fig. 1.6.** Structure of the SRC after A-120 is detected to be closed. The completion of the delivery for room A-120 is handled as an opportunity. As soon as RHINO learns that A-120 is open, the plan for the request is inserted in the actual schedule

The plan adaptor for rescheduling also adds an additional plan adaptor **P-3** that monitors the assumptions underlying the schedule, that is that the rooms A-110, A-111, A-113, and A-120 are open. The SRC after this revision is shown in Figure 1.5. After RHINO has picked up the red letter and left room A-111, it notices that room A-120 has been closed in the meantime (event 3 in Figure 1.3). Because RHINO cannot complete the delivery of the red letter the corresponding command fails. This failure triggers the replanning plan adaptor **P-2** which transforms the completion of the delivery into an opportunity (see Figure 1.6). Thus, as soon as RHINO notices room A-120 to be open (event 6) it interrupts its current mission, completes the delivery of the red letter (event 7), and continues with the remaining jobs after the red letter has been delivered.

## 1.5 Previous Work

To put this book into context we have to briefly review other approaches to controlling everyday activity and plan-based control of robotic agents.

### 1.5.1 Descriptive Models of Everyday Activity

A number of researchers in the fields of Psychology, Philosophy, Economy, Cognitive Science, and Artificial Intelligence have proposed and studied various competing models for capturing the nature and essence of conducting everyday activity. Miller, Galanter, and Pribram (1960) propose a view in which people perform everyday activity by starting from a goal, choosing an appropriate plan, and then executing the plan in closed-loop control. The plan is controlling the sequence of operations that the agent is carrying out. This view is dominant in much of the AI planning research and the earlier work on the plan-based control of autonomous robots.

Suchman (1985) argues that attributing human action to the formulation and execution of plans is untenable. She argues that the inherent unpredictability of the world makes all activity fundamentally improvised and that long-term planning can only guide the activity. The execution of plan steps requires the adaptation of the plan step to the particular, immediate situation. This process of fitting can, in the worst case, involve arbitrary domain knowledge, reasoning, and supporting activity in the world.

Bratman (1987) proposes that activity consists largely in the adoption, refinement, and reconsideration of intentions and plans. Plans have two important characteristics. First, they are typically incomplete. Second, plans are stable: normally, one reasons about how to execute and fill in one's plan, but not, unless special circumstances arise, about whether to reject the plan in favor of some other. Because plans are stable, plans make the practical reasoning one has to do manageable by framing one's deliberations, and so restricting the number of options that need to be considered.

Our approach exhibits characteristics of all three approaches. As Miller, Galanter, and Pribram (1960) we view all of the robot's activities as produced by plans. Here, we share McDermott's (1992a) view, who observes that any robot is under the control of a program and that a plan is simply that part of the control program that the robot explicitly reasons about and manipulates. We agree with Suchman (1985) in that the successful conduct of everyday activity requires improvisation and adaptation to specific situations. We disagree with her in that this implies that the robot is only guided but not controlled by a plan. For example, Firby (1989) has shown that control programs can be made to exhibit improvisational and situation specific behavior by adding specific control structures to the behavior specification language. From Bratman (1987) we borrow the idea that plans are valuable resources for focusing execution time deliberation.

### 1.5.2 Computational Models of Everyday Activity

Besides these descriptive models of everyday activity there have also been a number of computational models for controlling everyday activity. In this section we only consider the ones that are directly related to our approach to plan-based control.

Sussman (1977) has developed the HACKER computer system as a computational model of skill acquisition. HACKER is a problem-solving system that has been applied to the problem of stacking blocks in the blocks world. Sussman considers a skill to be a set of procedures each of which solves some problem in its domain. Sussman's model of problem-solving works as follows: An agent first tries to analyze the problem it intends to solve. If it knows of a method to solve the problem, a *skill*, then it applies the skill. Otherwise it must construct a new method or plan by applying more general problem-solving strategies. In the construction of the new methods the agent tries to avoid known problems and pitfalls (Sussman, 1990). If the method works it is remembered for future use. If not, the failures are examined and analyzed in order to apply debugging methods for repairing the method. Often the analysis of the failure can also be classified and abstracted and be remembered as a pitfall. HACKER can thereby learn critics which inspect the plans for known generalized bugs. Unfortunately, the computational techniques implemented in HACKER make all the assumptions from classical planning and therefore the techniques cannot be transferred easily to the problem of controlling robotic agents.

Despite the differences in the assumptions underlying the computational problems we consider, I borrow several ideas from Sussman's seminal work. First, the idea that agents learn from experience plan schemata for the problems they are intended to solve. Second, the idea of heuristically constructing almost good plans first, and then correcting the plans by examining and analyzing the flaws they produce. The analysis of the bugs often provides valuable information that suggests how the plan should be modified to be good.[2]

Wilensky (1983) has developed PANDORA a knowledge-based, commonsense planner. The assumptions underlying the PANDORA system are that plans for common problems are stored as canned plans indexed by the goals they are intended to achieve. The overall plan fails to achieve all goals if the plans for the individual goals interfere negatively. Thus planning is an iterative process that performs the following steps. First, detect the goals that might arise either through the user commands or from the use of the proposed plans. Second, select a possible plan by either finding a stored plan that matches the current goals or propose a new plan using the system knowledge. The plans are then projected, mentally executed, in order to check for conditions that will not be met, to notice bad side effects, and to adopt additional

[2] Note, in the general case, the generation of plans through plan debugging is as complex as generating the plans by chaining together primitive actions (Nebel and Koehler, 1995).

goals that are needed to address discovered problems. In the last step, the plan evaluation step, the planner decides whether the plan is acceptable.

The main ideas that we take away from Wilensky's Thesis are that plans for everyday activities can be pasted together from the plans for the individual goals using fast heuristic methods. The remaining flaws of the plan can be detected by projecting the plan, that is, predicting how the world will change as the plan gets executed.

PRS (Procedural Reasoning System), which has been developed by Georgeff and Ingrand (1989), is a computational model of the practical reasoning approach introduced above. PRS enables control systems to perform real-time task management that exploits information received asynchronously from multiple information sources such as the robot's sensors. The main ideas proposed in this approach is that task and plan management operations are triggered by asynchronous belief changes and that plans are useful data structures to focus the execution time reasoning needed for plan management.

McDermott (1978) describes NASL a very generic and flexible problem-solving system that interleaves planning and execution. NASL is applied to the domain of automatic electronic circuit design. NASL's plan language provides powerful control structures that allow for the flexible activity and for different kinds of synchronization between actions. McDermott also introduces the concept of constraining plans (secondary plans or policies) that are performed successfully if the subplans within their scope are executed in a particular way ("get all the groceries without spending more then twenty USD"). These control structures and the process of carrying out plans specified in the NASL plan language has also been formalized by McDermott (1985b). Many of the control structures used by the plan language employed in our research are reminiscent of the NASL plan language.

XFRM (McDermott, 1992b) is an experimental transformational planning system that is embedded in a simulated robot with restricted and noisy sensors and imperfect effector control. XFRM reasons about and manipulates reactive plans, that is, plans whose steps are conditional on the results of sensing actions. The plans are written in RPL (Reactive Plan Language), which has also been developed by McDermott (1991). XFRM operates as follows: (1) Given a set of commands from a user, it retrieves from a library a set of default plans for carrying those commands out; (2) it projects the plans, yielding a set of time lines that correspond to possible execution scenarios; (3) it runs critics on the scenarios, yielding suggested plan transformations; (4) the most promising transformation is carried out, and the process repeats. In parallel with this planning activity, the best current plan is being executed. When the planner thinks it has a better plan, it tells the executor to stop work on the current plan and switch to the improved version. Beetz et al. (2001); Beetz (2001) has extended the XFRM planning system by incorporating declarative goals into the reactive plans, developing a language for concisely specifying transformations of reactive plans, and realizing more effective means for execution time planning.

The plan-based controllers described in this book are implemented using the representational, inferential, and plan manipulation means provided by the extended XFRM system.

## 1.6 Contributions

This book describes part of an enterprise in which we seek to investigate computational principles that allow for the realization of a new class of plan-based controller for autonomous robots (Beetz et al., 2000). These plan-based controllers are characterized by using plans that are formulated in a very rich behavior specification language and can therefore produce flexible and reliable activity. They have very realistic and detailed models of their plans and the behavior the plans might produce. They are able to apply fast inferences for plan management because they assume that the robot is to carry out everyday activity. Therefore the robot already has flexible and reliable routine plans for carrying out its daily standard tasks. Plan management only operates to resolve interactions between individual plans and to handle dynamically changing situations. Finally, the controller can learn the plans for its routine activities from experience.

We believe that the research described in this book brings at least three important technical contributions to the problem of plan-based control of robotic agents.

1. **A plan representation for robotic agents performing everyday activity**. We show how the different mechanisms of autonomous robots including physical action, perception, planning, and communication can be represented and integrated into symbolic high-level plans without abstracting away from the different ways how the mechanisms work (Beetz and Peters, 1998). We will see that this requires that the plan representation provides the *expressiveness* of modern event-driven and real-time programming languages including loops, program variables, processes, and subroutines as well as high-level constructs (interrupts, monitors) for synchronizing parallel actions.
   This does, however, not need to entail that the execution time plan management operations have to perform automatic programming tasks for arbitrarily complex programs. Rather the concurrent plans can be designed to allow for fast prediction and plan failure diagnosis, transparent access to the relevant plan pieces, and to enable successful continuations after revisions and interruptions. We support the fast and reliable execution of plan management operations by identifying common control patterns and encapsulating them into task-specific sub-plan macros, which represent complex control fragments modularly and transparently.

2. **Advanced inference mechanisms for plan-based control of robotic agents.** The first contribution in the area of inference mechanisms for robot planning are *Probabilistic Hybrid Action Models* (PHAMs) (Beetz and Grosskreutz, 2000), a realistic causal model for predicting the behavior generated by modern concurrent percept-driven robot plans. PHAMs represent aspects of robot behavior that cannot be represented by most action models used in AI planning: the temporal structure of continuous control processes, their non-deterministic effects, and several modes of their interferences. We will give a formalization of PHAMs and proofs that the formalized model generates probably, qualitatively accurate predictions. In addition, we describe a resource-efficient inference method for PHAMs based on sampling projections from probabilistic action models and state descriptions (Beetz and Grosskreutz, 1998).
   The second contribution in the area of inference mechanisms for robot planning is *Probabilistic Prediction-based Schedule Debugging* (PPSD) (Beetz, Bennewitz, and Grosskreutz, 1999), a novel planning technique that enables autonomous robots to impose order constraints on *concurrent percept-driven plans* to increase the plans' efficiency. The basic idea is to generate a schedule under simplified conditions and then to iteratively detect, diagnose, and eliminate behavior flaws caused by the schedule based on a small number of randomly sampled symbolic execution scenarios.
3. **Acquisition of symbolic robot plans.** We will describe XFRMLEARN, a system that learns symbolic structured plans for standard navigation tasks (Beetz and Belker, 2000b). Given a navigation task, XFRMLEARN learns to structure continuous navigation behavior and represents the learned structure as compact and transparent plans. The structured plans are obtained by starting with monolithic default plans that are optimized for average performance and adding sub-plans to improve the navigation performance for the given task. Compactness is achieved by incorporating only sub-plans that achieve significant performance gains (Beetz and Belker, 2000a). The resulting plans support action planning and opportunistic task execution.

As evidence of the feasibility of our approach to plan-based control of robotic agents and of the performance gains that might be expected from its application, we have performed several longterm demonstrations. The first demonstration has been an SRC controlling the high-level course of action of a museum tour-guide robot for thirteen days and showed the flexibility and reliability of runtime plan management and plan transformation. During its period of operation, it was in service for more than 94 hours and made about 3200 plan transformations while performing its tour-guide job (Thrun et al., 1999; Thrun et al., 2000). In another experiment we have evaluated more specific capabilities of the plan-based controller. In one experiment, we have shown how predictive plan transformation can improve the performance by outperforming controllers without predictive transformations in situations that require foresight while at the same time retaining its performance in situations that require no foresight (Beetz, 2001).

## 1.7 Overview

In the remainder of this book we proceed as follows. In the next chapter, we begin our discussion with basic models that will be used throughout the book. We introduce the dynamic system and the rational agent model of autonomous robot control. In this chapter we will also give an overview of the overall control system, in which the plan-based control is integrated. Finally, the chapter describes the structure and the operational principles of the high-level controller.

The following chapter describes the self-adapting plans and our plan representation for the robot's primary activities. We start with explaining the integration of the robot's mechanisms into the plan language. After the description of the low-level plans we will then discuss how plans for delivery tours can be represented transparently and modularly. At the end of the chapter we will explain how the plan management operations are specified.

The fourth chapter describes a causal model of the behavior exhibited by a mobile robot when running concurrent reactive plans. The model represents and predict aspects of robot behavior that cannot be represented by other action models used in AI planning: it represents the temporal structure of continuous control processes, several modes of their interferences, and various kinds of uncertainty. We will detail the causal models and show that the models can, with high probability, make quite accurate predictions. The last part is about how the prediction mechanism can be implemented to economize computational resources.

The problem of learning symbolic plans that support action planning and opportunistic task execution is addressed in the fifth chapter. We will show in the context of learning structured reactive navigation plans that the robot can automatically learn to structure continuous navigation behavior and represent the learned structure as compact and transparent plans. We will explain how these plans can be obtained by starting with monolithic default plans that are optimized for average performance and adding subplans to improve the navigation performance for the given task.

Chapter 6 describes the field tests and the feasibility studies that we have performed to evaluate our approach to plan-based control.

We conclude with an evaluation and a discussion of related work.

# 2. Overview of the Control System

This chapter describes our conceptualization of, and a software architecture for, plan-based control of robotic agents. The chapter is organized as follows. We will start in section 2.1 with describing an abstract control model for the class of robot control problems that we investigate. This model integrates aspects of a dynamic system model, which is primarily used in control theory, with a "rational" agent model, that is primarily used for the characterization of problem-solving systems in Artificial Intelligence. Section 2.1 also introduces key concepts and terminology used in the remainder of this Thesis. Section 2.2 describes different environment representations that support in particular the plan-based control of the robotic agent. Section 2.3 outlines the computational structure of the overall control system in which our plan-based control mechanisms are integrated.

## 2.1 Abstract Models of Robotic Agents

An important prerequisite of the successful investigation of control methods for autonomous robots is the use of appropriate models of the robot, its environment, its perception of the environment, and the way control is exerted. In this section we will introduce such a model that integrates aspects of the dynamic system model and the rational agent model. The dynamic system model is used for the conceptualization of the interaction between the robot and its environment. The rational agent model is concerned with how the robot chooses the right actions to perform.

### 2.1.1 The Dynamic System Model

In the dynamic system model, the state of the world evolves through the interaction of two processes: one denoted the controlling process — the robot's control system — and the other the controlled process, which comprises events in the environment, physical movements of the robot and sensing operations.[1] The physical objects involved in the controlled process are the

[1] Dean and Wellmann (1991) and Passino and Antsaklis (1989) give very informative overviews of the use of dynamic system models in AI planning.

hardware components of the robot and the objects in the robot's operating environment. The purpose of the controlling process is to monitor and influence the evolution of the controlled process so that the controlled process satisfies the given constraints and meets the specified objectives. Using the terminology of control theory, we will denote the controlled process as the *plant* and the controlling process as the *controller*.

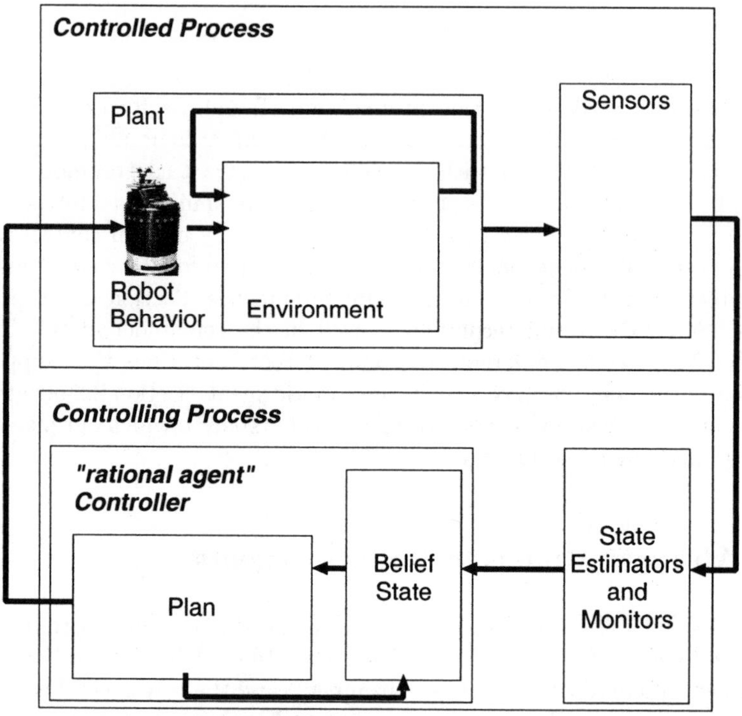


**Fig. 2.1.** Block diagram of our dynamic system model for autonomous robot control. Processes are depicted as boxes and interactions as arcs. The dynamic system consists of the controller (or controlling process) and the plant (or controlled process). The plant is decomposed into the environment and sensor process and the controller into the state estimation and the ration agent process

Figure 2.1 shows a block diagram of the dynamic system model that underlies the design of our robot control system. The processes are depicted as boxes and the interactions between them as arcs. There are two interactions between the controlling and the controlled process. The first interaction is initiated by the controlling process that sends control signals to the controlled process in order to influence the evolution of the controlled process in the desired way. In general, the controlled process is only under partial control of the controlling process. Concurrently with the stream of control signals events that are not under the control of the controller occur and change the state of the controlled process, too. The second interaction consists of the controlling process observing the controlled process by interpreting the sensor data that are produced by the controlled process.

For autonomous robot control, it is often useful to decompose the controlled process into an environment and a sensing process (Dean and Wellmann, 1991). The environment process comprises the actions of the robot, which are primarily different kinds of movements and communication actions, that cause changes of the environment state. The robot concurrently performs multiple actions that might interfere. For example the effects of different movements might interfere through transposition. The subprocesses of the environment process have discrete as well as continuous effects. When the robot is navigating through the environment, its position is changing continuously. The actions of the robot are not the only processes that change the environment. The environment also changes due to the occurrence of exogenous events such as the opening and closing of doors and due to physical processes such as the coffee getting colder while it is delivered.

The sensing process maps the world state into the sensor data, the input of the controlling process. Sensors can provide continuous streams as well as individual bits of sensor data. It is important to note that the robot's sensors can, in general, access only partial information about the state of the environment. The sensor data measured might also be inaccurate and must therefore often be interpreted to be useful for controlling the robot. The inaccuracy, locality of sensor data as well as the data interpretation processes yield data that are incomplete, inaccurate, and ambiguous.

The decomposition of the controlled process suggests an analogous decomposition of the controlling process into a state estimation and a "rational agent" process. For now, we will consider the rational agent process to be a black box that continually receives percepts generated by the state estimation processes and outputs control signals for the controlled process. We will discuss what the rational agent should do and how it should be designed in the next section. The state estimation processes compute the robot's beliefs about the state of the controlled system. The monitoring processes signal system states for which the controlling process is waiting. The results of the state estimation and monitoring processes are stored in the belief state of the robot controller. The belief state contains information such as the robot's estimated position, the accuracy and a measure of the ambiguity of the position estimate, the state of objects to be delivered, and opening angles of doors. It is important to note that state estimation and monitoring processes are often passive and run continuously. As a consequence, the belief state changes asynchronously with the robot's own actions.

The agent controller specifies the control signals that are supplied to the controlled process as a response to the estimated system state. At the physical level most inputs generated by the controlling process are voltages for the different motors of the robot. At a higher level of abstraction the control signals are realized by a more abstract programming machine that provides the control system's application programmer's interface.

### 2.1.2 Autonomous Robots as Rational Agents

Let us now turn to the question of what the rational agent process is supposed to do and how it can be implemented. The reason that we realize the robot controller through a so-called rational agent process instead of an ordinary control program is that for complex and dynamically changing jobs and complex environments it is often impossible to think through all these decisions in advance or specify them concisely. For such applications it is often more promising to equip the robot controller with decision criteria and let the robot decide at execution time based on its assessment of the immediate situation. Such an approach is best described using the abstract model of a "rational agent."

**The General Agent Model.** Russell and Norvig (1995) consider agents to be systems that they want to characterize in terms of their perceptions and actions. Agents interact with their environment by means of sensing their environment and acting in response to these percepts in order to change environment's state. The agent is carrying out a feedback loop in which it selects the next action based on the sequence of percepts that it has received so far. The execution of the action produces behavior which in turn changes the state of the environment. The sensors of the agent then produce the percepts that are used for the action selection in the next cycle of the agent's operation. Thus, the agent's operation in the environment is adequately described as an *inter*action with the environment.

In this model, agents and environments are defined in terms of the actions and percepts that they exchange and the states that they traverse. What an agent does can be described abstractly by a function, the so-called *agent function*, that maps sequences of percepts into an action chosen to be executed. In designing the controller of an autonomous robotic agent, the key problem is to realize agent functions that choose the "right" actions quickly enough.

Before we can state what the right actions are, we must first establish how we measure the performance of the agent. An external observer can evaluate the performance of an agent by recording the sequence of states that the environment traverses while the robot performed a problem-solving episode and mapping the state sequence using a measurement function into a value that represents how well the agent did in the corresponding problem-solving episode. This process of measuring the performance of an agent is depicted in Figure 2.2.

**Rational Agency.** Now that we know how to measure the performance of an agent, we are in the position to state what we mean by an agent being *rational*. A *rational agent* is an agent that attempts to maximize its

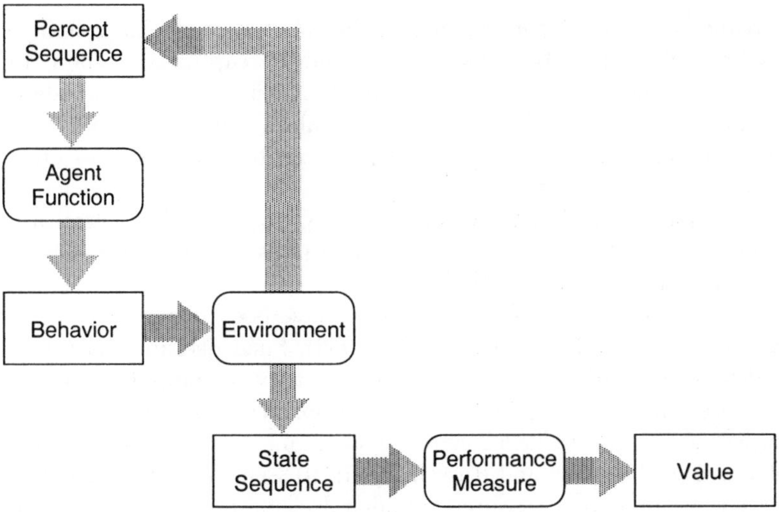


**Fig. 2.2.** The figure shows the interaction of an agent, characterized by its agent function, with its environment. The performance of the agent is measured by an external observer using a function that maps the state sequence that the environment process traverses into a numerical value

expected performance in the eyes of an external observer. The agent does so by making informed action selections on the basis of its percepts and its knowledge. Thus, an agent function is rational, if there exists no other agent function that can be expected to produce better performance with respect to the beliefs of the agent. In the case where the agent has to select its actions based on uncertain information, decision theory provides the mathematical foundation for the realization of the agent function (Ramsey, 1931; Haddawy and Rendell, 1990). Using decision-theoretic action selection, the best action is chosen using utility functions that map state sequences into numerical measures. Utility functions are subjective to the agent and also encode the agent's attitude towards risks, whether the agent is risk aversing or risk seeking (Koenig and Simmons, 1994; Haddawy and Hanks, 1992). Note, the expectation is taken according to the agent's own beliefs; thus, perfect rationality does not require omniscience.

**Resource-Bounded Rationality.** Recall that agent functions have to supply the robot with a chosen action in each perception-action cycle. On the other hand, executing the agent function on the robot's computer, which only provides limited computational resources, might yield the controller arriving at a decision after the decision is needed. If no action is chosen, the robot must either wait or execute a default action, both of which typically reduce the performance of the robot. As a consequence, the computational resources required to select an optimal action may reduce the overall performance of the robotic agent (Zilberstein, 1996; Russell and Wefald, 1991). This state of affairs is captured by the notion of resource-bounded rationality (Simon, 1982). The notion of resource-bounded rationality is stricter than that of computability and complexity because resource-bounded rationality relates the operation of a program on a machine model with finite speed to the actual temporal behavior generated by the agent (Russell and Subramanian, 1995). The goal is the optimization of computational utility given a set of assumptions about expected problems and constraints on resources (Horvitz, 1988).

Resource-adaptive decision procedures have been proposed to deal with these requirements (Dean and Boddy, 1988). Such decision procedures return default or heuristic solutions quickly and try to compute better results given more computational resources. The problem of assigning computational resources to composite decision processes requires sophisticated reasoning mechanisms (Zilberstein and Russell, 1995). Russell and Subramanian (1995) observe that competent resource-bounded decision procedures for stochastic deadlines often require more complex program structures.

### 2.1.3 The BDI Model of Rational Agents

The BDI (belief, desire, intention) model is a specific framework for the realization of the agent functions of bounded rational agents (Rao and Georgeff, 1992). The BDI model is tailored towards agents that act in dynamic and uncertain environments and are limited with respect to the information that they can access and the computational resources they can consume. The BDI controller receives percepts asynchronously and outputs streams of control signals (Georgeff and Ingrand, 1989).

To deal with these characteristics the computational state of a BDI controller is composed of three main data structures: the beliefs, goals, and plans of the agent. The beliefs comprise information about the environment, the state of the environment, the agent state, and the actions that the agent may perform. Beliefs are essential because the world is dynamic and the system only has a partial view of the world. Moreover, as the system is resource bounded, it is advantageous to cache important information rather than recompute it from base perceptual data over and over again. The goals represent some desired end or intermediate states and thereby give the agent important means to recover from failures and find alternative means for accomplishing the agent's jobs. BDI architectures are plan-based systems that use plan libraries with plan schemata for achieving particular goals. At execution time plans are pasted together and filled out with sub-plans based on the actual state of the world. The plans are the courses of actions that the agent intends to perform. Plans are used to focus the reasoning of the agent by filtering out options that conflict with the intended plans.

The content of the belief and goal data structures of a BDI agent determines how events and goals give rise to plans and how plans lead to action and the revision of beliefs and goals. The mechanisms of a BDI agent ensure that beliefs, goals, and intentions evolve rationally. That is, that events are responded to in a timely manner, that beliefs are maintained consistently, and that the plan selection and execution proceeds in a manner which reflects certain notions of rational commitment.

Bratman, Israel, and Pollack (1988) have developed IRMA (Intelligent Resource-Bounded Machine Architecture) as a specific realization of a BDI architecture. The key idea in IRMA is that agents should, in general, bypass full-fledged deliberation about new options that conflict with their existing commitments, unless these options can easily be recognized as potentially special in some way. To achieve this balance between commitment to existing plans and sensitivity to particularly important new options, Bratman, Israel, and Pollack (1988) propose a filtering mechanism with two components. The first checks the compatibility of a new option against existing plans, and the second checks whether the new option is important enough to warrant deliberation even if already intended sub-plans must be suspended or revised if the new option is accepted.

The Procedural Reasoning System (PRS) is another implementation of a BDI architecture that is specifically designed for resource-bounded, goal-directed, and situated control (Ingrand, Georgeff, and Rao, 1992; Georgeff and Ingrand, 1989). The architecture enables control systems to perform real-time task management that exploits information received asynchronously from multiple information sources such as the robot's sensors.

PRS has been successfully applied to very complex control problems including space shuttle fault diagnosis (Georgeff and Ingrand, 1990), mobile robot control (Myers, 1996), air traffic control, and air battle management.

### 2.1.4 Discussion of Our Robotic Agent Model

Our robotic agent can be characterized as a bounded rational agent that acts as the controller in a dynamic system. The controlled process is composed of concurrent, temporally extended, subprocesses that have continuous, possibly interfering, effects. The behavior of the controlled process is best described as trajectories in the state space that is spanned by the state variables of the system. The controlled process is only partially observable and the observations of the observable part are imperfect. The inputs for the rational agent process is generated by the state estimation and monitoring processes and are received asynchronously with respect to the controller's own operation.

The controller is realized as a BDI agent where the beliefs concerning the state of the controlled process are computed by the state estimation processes. The task of the controller is to manage a plan that is consistent with the robot's beliefs and the user requests. Given that situations which require reactions from the robot controller can arise asynchronously with respect to the controller's operation, control routines are made robust by incorporating the monitoring of actions and by reactions triggered by observed events.

The main difference between the research conducted in this Thesis and the research conducted on the BDI systems discussed before is that the designers of the PRS system, for example, interpret the notion of planning fairly general. In their view planning is any choice as to which course of action to pursue, no matter how the system does it. In this Thesis the focus is on developing plan representations that are specifically designed to facilitate execution time plan management. We also develop powerful inference mechanisms for difficult reasoning tasks, such as predicting the behavior generated by concurrent reactive plans realistically, that should make prediction-based execution time plan management feasible.

## 2.2 The Environment Maps

Many plan-based robot controllers use maps, models of their operating environments, as a resource for performing their tasks more reliably and efficiently. In general, the kinds of information stored in robot maps and the way information is organized is tailored for the robot's tasks. A cleaning robot, for instance, profits from a map that allows for fast retrieval of the regions to be cleaned and that supports the generation of optimal tours that cover the area to be cleaned. A tour-guide robot, on the other hand, should use maps that support self-localization even when the environment is crowded with people and that contain obstacles that cannot be detected by the robot's sensors.

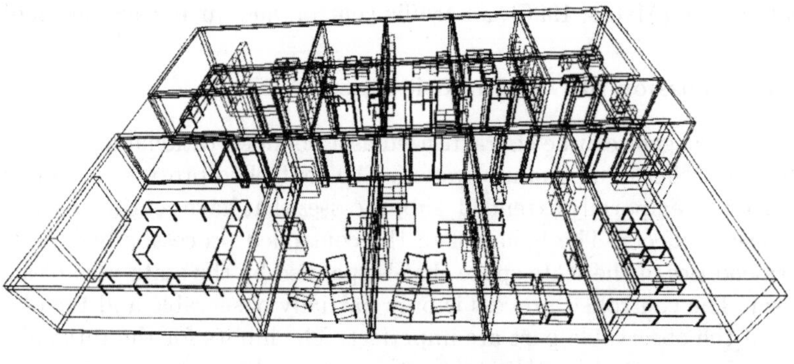

**Fig. 2.3.** Environment map used by RHINO's plan-based controller. The environment map contains the walls, doors, and the pieces of furniture that are assumed to be static. The items modeled in the map have a hierarchically structured geometric model and a symbolic description associated with them

RHINO's plan-based controller uses a *symbolically-annotated 3D map* (SATD map) (Beetz et al., 1999), which enables RHINO to perform tasks such as "get a book from the shelf in room A-120 and bring it to Wolfram's desk."

To provide the necessary information SATD maps store models of the task-relevant objects such as desks and shelves. The task-relevant objects themselves are represented as hierarchically structured 3D models, which enables RHINO to identify the part of the camera image that corresponds to the table top of a desk. Object models can be retrieved through relational symbolic queries such as "the desk in room A-120." Finally, aspects of the environment structure are represented explicitly. This enables RHINO to adapt its behavior to the different parts in the environment (different driving strategies in offices and hallways, etc) and detect problems such as closed doors.

An SATD map consists of a set of object models and a set of structural elements. An object model has a name and consists of a specification of its category (desk, shelf, table, ...), a geometric model, the position of the center of the object and its orientation, and a set of attribute-value pairs of the form {⟨category,desk⟩, ⟨color,red⟩, ...}. Structural elements, such as smoking areas or rooms, represent regions of the environment. With each region is associated a set of doorways through which the region can be entered and left (cf. (Kortenkamp and Weymouth, 1994)).

Figure 2.3 shows a SATD map of part of our institute. The query

select d where d.category = desk ∧ d.room = A-120

retrieves all desks in room A-120. The result of the query is shown in Fig. 2.4.

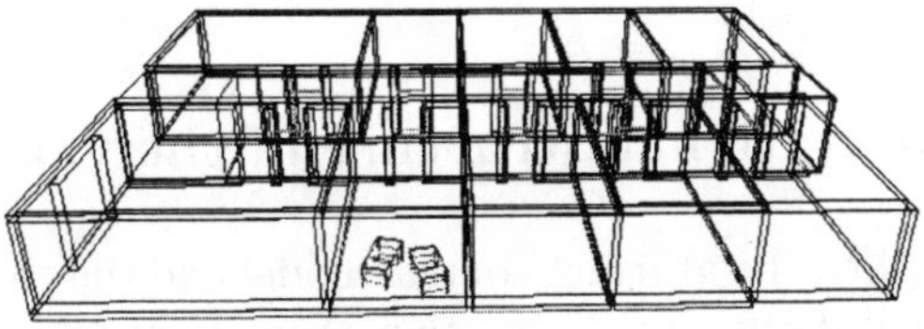

**Fig. 2.4.** Result of a query to a symbolically annotated 3D map. The figure shows the items in the SATD map that satisfy the query select d where d.category = desk ∧ d.room = A-120

SATD maps improve RHINO's problem solving competence in various ways. For example, RHINO uses SATD maps for interpreting locational descriptions of objects to be searched. The locational descriptions are then translated into probability distributions for the locations of objects. Figure 2.5(b) shows a probability distribution for the location of a letter given that the letter is probably close to a desk.

SATD maps can also be used to generate occupancy grid maps for robot self-localization. These grid maps are generated by making a projection of the 3D map at a given height (e.g., the height of the robot's sonar sensors)

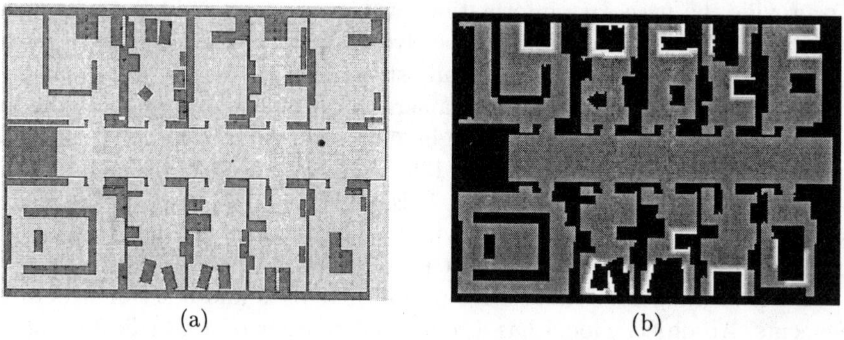

Fig. 2.5. Two task-specific environment maps. (a) An occupancy grid map in table height for self localization and navigation planning. (b) An object probability grid map that represents the probability distribution for an object's location for informed object search. The brighter a grid cell the more likely it is that the target object is at this position

(see figure 2.5(a)). These maps are used by a localization method (Fox et al., 1999) to estimate the robot's position with an average accuracy of 5-10cm. Other uses of the SATD maps include simplifications of place recognition and restricting visual search to certain regions of the image, such as the top of a table. Or, to enable the robot to perceive events such as passing a door or entering the hallway (Beetz et al., 1998).

## 2.3 The Computational Structure of the Control System

After having described the abstract control model and the environment maps that we use, we will discuss in this section an orthogonal issue: the computational structure of our control system.

Figure 2.6 depicts the computational structure of the variant of RHINO-control system that is used in this Thesis. The control system is composed of two layers: the *functional* and the *task-oriented layer*. At the functional layer the system is decomposed into distributed modules where each module provides a particular function (cf. (Alami et al., 1998)). A function could be the control over a particular actuator or effector, access to the data provided by a particular sensor subsystem, or a specific function of the robot's operation, such as guarded motion. The functional layer is decomposed into the perception and the action subsystem. The modules of the perception subsystem perceives states of the environment and the modules of the action subsystem control the actuators of the robot.

The task-oriented layer is constituted by the Structured Reactive Controller that specifies how the functions provided by the functional layer are to be coordinated in order to perform specific jobs such as the delivery of

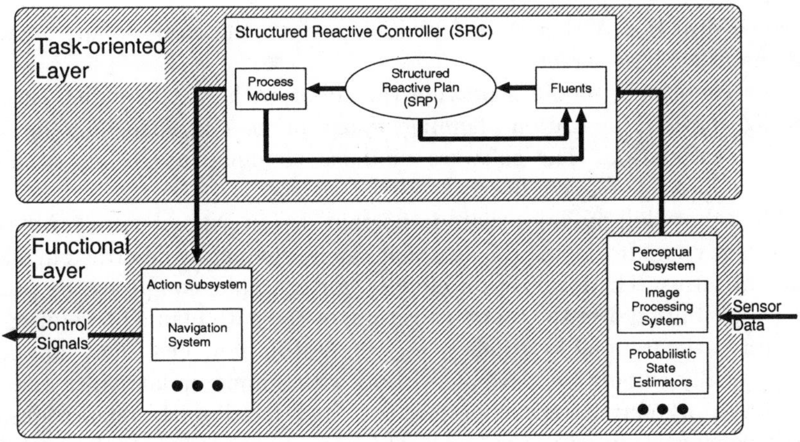


**Fig. 2.6.** The computational structure of the RHINO control system. The figure depicts the two layers of the system. We consider the functional layer as being decomposed into a perception and an action subsystem. The task-oriented layer is constituted by the Structured Reactive Controller. The perception subsystem computes the information for setting the fluents of the SRC. The process modules that are activated and deactivated by the Structured Reactive Plan control the modules of the action subsystem

objects. The perception subsystem computes the information for setting the fluents of the SRC. The process modules that are activated and deactivated by the Structured Reactive Plan control the modules of the action subsystem.

In the remainder of this section, we will proceed as follows. Section 2.3.1 describes the main modules at the functional layer at a greater detail. Section 2.3.2 describes the abstract machine that constitutes the interface between the functional layer and the Structured Reactive Plan, which is sketched in Section 2.3.3.

### 2.3.1 The Functional Layer

The functional layer of the RHINO system consists of approximately 20 distributed modules, each of which is designed to monitor or control dedicated aspects of the robot and additionally to provide information to other modules. Each module is capable of performing a number of different but related functions. The modules at the functional layer provide the basic means for moving the robot, sensing the surroundings of the robot, interacting with humans (using the robot's buttons), etc.

We distinguish between *server modules* and *robot skill modules.* Server modules are associated with a particular device. The camera server, for example, enables the robot to request an image of a particular type and obtain the image. The pantilt server allows the robot to point the camera into a specified direction. Other server modules include the laser, sonar, buttons, and the robot base server.

The robot skill modules provide the robot with specific skills, such as guarded motion, navigation planning, image processing, map building, and position estimation. The software modules communicate using an asynchronous message-passing library.

The robot skill modules share important design characteristics. Many of them solve computational problems that can be compactly described in mathematical terms. The localization module, for example, computes the probability distribution over the robot's position, or the path planner computes the shortest paths from all locations to a given destination. The robot skill modules typically employ iterative algorithms with simple update rules and can therefore run many iterations per second. Such fast iterative algorithms are particularly well suited to the low-level control system because they are able to react almost immediately, can promptly accommodate asynchronously arriving information and often perform anytime computations which produce initial results very quickly and can continue to improve upon these results given more computation time.

The functional layer also employs HLI (High-level Interface) (Hähnel, Burgard, and Lakemeyer, 1998) as a module that monitors the operational status of the modules at the functional level, resets and reconnects modules after module failures, and provides some basic means for performing non-trivial tasks more more reliably. It does so by synchronizing the operations at the functional layer in task specific ways and by recovering from local execution failures.

**Perceptual Subsystems.** RHINO's perceptual subsystem contains state estimation, image processing, button interaction, and email server modules.

*Probabilistic State Estimators.* Taking inaccurate, local, and incomplete sensor measurements as input, RHINO's state estimators provide abstract percepts to the high-level system. Examples of such abstract percepts are opening angles of doors, the position of the robot, inaccuracy and ambiguity of the perceived position, etc. Other RHINO modules can register for updates of abstract percepts, for example of a door which needs to be passed, or of the position of a wastebasket for which the robot is looking. The state estimators then check for this information and send updates to the high-level system every time a change is detected.

A key prerequisite for effective and efficient navigation is that the robot knows its own position within its environment. To maintain a position estimate of its position the controller concurrently executes a particular state estimation process: the localization process (Fox et al., 1999). The estimate computed and maintained by the localization process is represented as a probability distribution over the possible positions of the robot. This distribution is updated continually using the motion commands that the controller issues and comparing the measured distances to the next obstacles to those that can

**Fig. 2.7.** The figure shows a probability distribution over the robot's position in the environment. The darkness of a position represents the probability that the robot is at this position.

be expected based on the environment map. Figure 2.7 shows a probability distribution over the robot's position in the environment. The darkness of a position represents the probability that the robot is at this position.

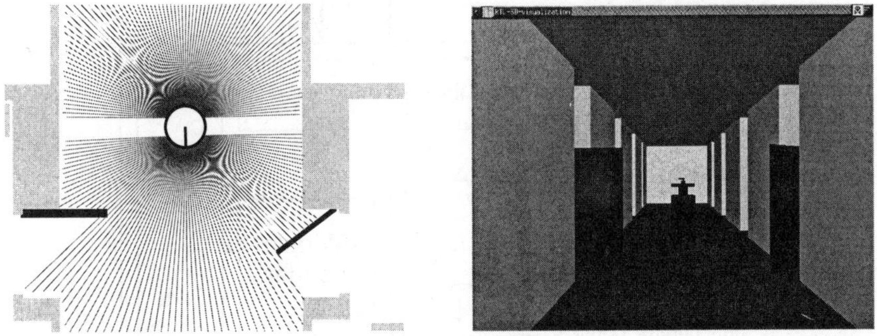

(a) (b)

**Fig. 2.8.** This figure pictures the door state estimation process. (a) Topview of the robot and the laser distance readings that it receives. (b) Visualization of the estimated opening angles of the doors. (Source: (Schulz and Burgard, 2001)).

The probability distribution is abstracted into three state variables: the robot's position and orientation, the accuracy of the robot's position estimate (the diameter of the global maximum of the probability distribution), and the ambiguity (the number of local maxima that exceed a certain probability threshold).

For the robot office courier we use an additional state estimator, which estimates the state of objects, based on the information provided by 180 degree laser range scanners. This state estimator handles objects which are permanent parts of the dynamic world model and which are subject to infrequent state changes like doors, chairs, wastebaskets and so on (Schulz and Burgard, 2001). The state density of these objects is estimated using a template matching approach which compares the real sensor measurements against ideal measurements for each sample state.

*Image Processing Subsystems.* The RHINO system's visual and image processing components comprise a system module and a programming framework called RECIPE: Reconfigurable, Extensible, Capture and Image Processing Environment (Arbuckle and Beetz, 1999 and 1998). As a programming framework RECIPE provides a standardized, multi-threaded, environment in which modules for image processing are dynamically loaded and unloaded. The framework provides standardized facilities for resource management and image capture to improve system robustness and to promote code re-use. RECIPE modules, as well as being run-time loadable, are scriptable, active, entities.

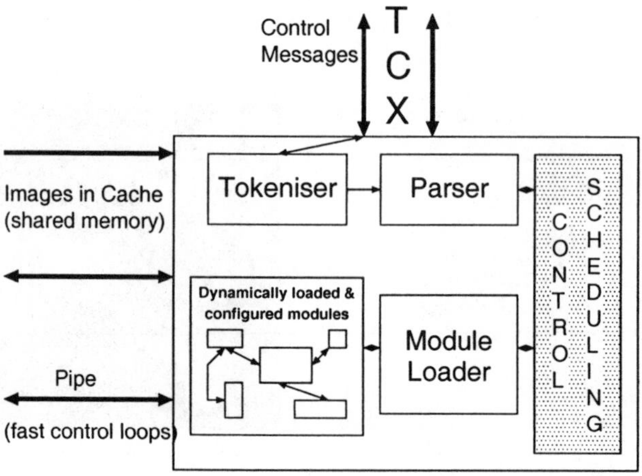


**Fig. 2.9.** Software architecture of RECIPE's image processing server. Control messages sent by remote modules are parsed and processed by the control scheduler, which loads and unloads special purpose image processing modules and passes requests to the responsible module

As a system module, RECIPE is responsible for the capture and processing of the visual information controlled. It has two principal components — a capture server and several copies of an image processing server. The capture server is a lightweight, robust process which has three main responsibilities. It stores the images captured by the system's cameras. It also stores all important system data and configuration information for RECIPE. The capture server acts as a watchdog for all image processing servers, restarting and reconfiguring them if their processing should cause them to hang or crash.

The image processing server forms the heart of the RECIPE system. It is a script controlled modular architecture for the implementation of image processing operations and pipelines. The image processing server loads and unloads image processing functionality according to commands sent by the remote modules, for example the plan-based robot controller. An image processing server is shown in figure 2.9.

*Email Server.* To provide natural language communication capabilities the RHINO system has been extended by two additional modules: the "email server" and the "natural language server." The "email server" broadcasts each electronic mail sent to the computer account "RHINO" to the modules that have subscribed to "automatic" email update. It also provides the functions necessary for sending electronic mails. The "natural language server" provides the functions required for transforming electronic mails into an internal representation and analyzing the content of the email using the RHINO's world model.

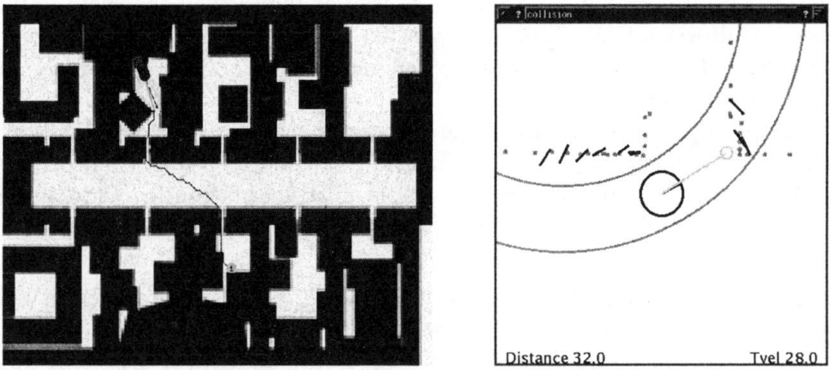


(a) (b)

**Fig. 2.10.** RHINO's navigation system consists of a navigation path planner (a) and a reactive collision avoidance module (b). Subfigure (a) shows an intended path. Subfigure (b) shows the laser and sonar readings and a circular trajectory computed by the collision avoidance module that will not cause any collision within the next two seconds.

**Navigation System.** The navigation system as currently used is described in (Thrun et al., 1998). It is a safe and reliable high-speed navigation system which has shown impressive results in several longterm experiments (e.g., (Thrun et al., 1999)). Conceptually, the RHINO navigation system works as follows. A navigation task is transformed into a Markov decision problem and solved by a path planner using a value iteration algorithm. The solution is a policy that maps every possible location into the optimal heading to reach the target. This policy is then given to a reactive collision avoidance module that executes the policy taking the actual sensor readings (and thereby unmodeled obstacles) and the dynamics of the robot into account (Fox, Burgard, and Thrun, 1997).

### 2.3.2 The "Robotic Agent" Abstract Machine

While the functional layer provides the basic functionalities of the robot control system – such as localization and navigation methods – in the form of continuous control processes that can be activated and deactivated, it does not provide effective means for combining the processes into coherent task-directed behavior. Without additional functionality, therefore, it would be tedious and error-prone to specify reliable and efficient control programs that were to perform non-trivial and composed control tasks.

This additional functionality is provided by the "Robotic Agent" abstract machine. The heart of the "Robotic Agent" abstract machine is constituted by RPL (Reactive Plan Language) (McDermott, 1991) the language in which the robot's plans are specified. RPL provides conditionals, loops, program variables, processes, and subroutines as well as high-level constructs (interrupts, monitors) for synchronizing parallel actions. To make plans reactive and robust, it incorporates sensing and monitoring actions, and reactions triggered by observed events.

The primitive components of the abstract machine are percepts and process modules, which form a uniform interface to continuous control processes. In addition to the primitive components the abstract machine provides mechanisms for combining percepts into more specific percepts and control structures for combining control processes into more sophisticated ones.

**Connecting Control Routines to "Sensors".** Successful interaction with the environment requires robots to respond to events and to asynchronously process sensor data and feedback arriving from the control processes. RPL provides *fluents,* registers or program variables that signal changes of their values. In order to trigger control processes fluents are often set to the truth value true in one interpreter cycle and reset in the following one. We call this setting and resetting of fluents pulsing. Fluents are also used to store events, sensor reports and feedback generated by low-level control modules. Moreover, since fluents can be set by sensing processes, physical control routines or by assignment statements, they are also used to trigger and guard the execution of high-level control routines.

For example, the robot plans use fluents to store the robot's estimated position and the confidence in its position estimation. Fluents can also be combined into digital circuits that compute derived events or states such as being in the hallway. For example, figure 2.11 shows a fluent network, which has the output fluent *IN-HALLWAY?* that is true if and only if the y-coordinate of the robot's global position larger than 780cm and smaller than 1330cm.

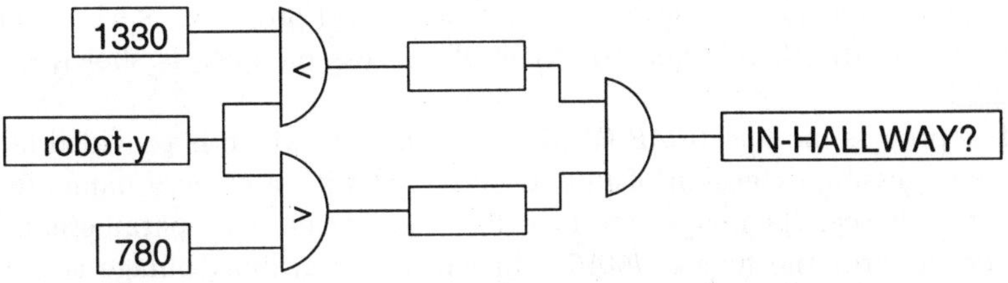


**Fig. 2.11.** Fluent network computing the percept "leaving a room". The fluent network receives the y- coordinates of the robot's estimated position as its inputs. The output of the fluent network is true, if and only if the value of the estimated y-position lies between 780 and 1330 cm, that is if the robot is in the hallway

Fluents are best understood in conjunction with the RPL statements that respond to changes of fluent values. The RPL statement *whenever* $F$ $B$ is an endless loop that executes $B$ whenever the fluent $F$ gets the value "true." Besides *whenever* , *wait for* ($F$) is another control abstraction that makes use of fluents. It blocks a thread of control until $F$ becomes true.

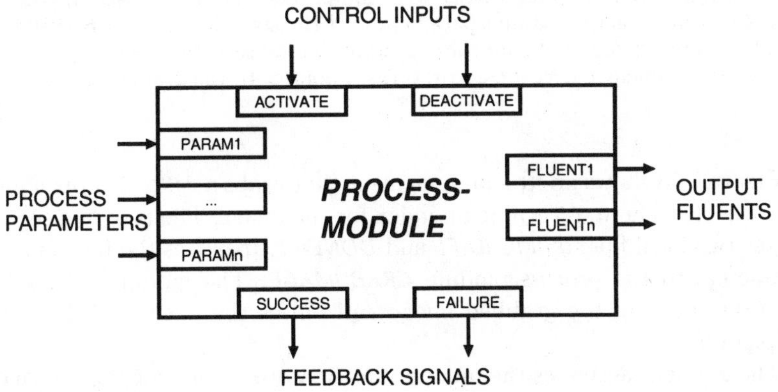


**Fig. 2.12.** Process modules provide an encapsulation for control processes with a uniform interface that can be used to monitor and control continuous control processes. Control processes can be activated and deactivated and return upon their termination success and failure signals. They also allow the high-level plans to monitor the progress of their execution by updating fluents that can be read from the plan

**Process Modules.** To facilitate the interaction between control routines and continuous control processes, the abstract machine supports a uniform protocol (cf. (McDermott, 1992b; Firby, 1994)). The control abstraction that implements this protocol is called a *process module*. Process modules are elementary program units that constitute a uniform interface between "high-level" plans and the "low-level" continuous control processes, such as image processing routines. A schematic view of process modules is shown in figure 2.12.

The process module *GRAB-IMAGE*, for example, which is provided in our image processing extension of RPL is activated with a camera name, image size, and whether the image is to be color or grayscale as its parameters. The module updates the fluents *IMAGE*, in which the grabbed image is stored, and *DONE?*, which is pulsed upon process completion.

```
with local fluents (IMG-ID-FL DONE-FL)
  GRAB-IMAGE(:CAMERA           :LEFT
             :SIZE             (240 320)
             :COLOR            :TRUE
             :GRABBED-IMAGE IMG-ID-FL
             :DONE             DONE-FL)
  wait for (DONE-FL)
  FLUENT-VALUE(IMG-ID-FL)
```

**Fig. 2.13.** RPL plan fragment for capturing an image. The plan fragment creates two local fluents for communicating with the process module *GRAB-IMAGE*. The process module is activated and parameterized through the call of *GRAB-IMAGE*. The plan fragment then waits for the control process to terminate, which is indicated by pulsing the fluent *DONE-FL* and returns a pointer to the grabbed image

Figure 2.13 shows an RPL plan segment that grabs a *240 × 320* pixel image with the left camera, returning an index for the image. The program segment creates two local fluents *IMG-ID-FL* and *DONE-FL*. It passes the fluents as call parameters to the process module *GRAB-IMAGE*. The module sets *IMG-ID-FL* to the index of the grabbed image and pulses the fluent *DONE-FL* upon completion.

The segment activates the module *GRAB-IMAGE*, waits for the image capture to complete and returns the value of the fluent *IMAGE-ID-FL*.

**Control Process Composition.** The abstract machine provides RPL's control structures for reacting to asynchronous events, coordinating concurrent control processes, and using feedback from control processes to make the behavior robust and efficient (McDermott, 1991).

Figure 2.14 lists several control structures that are provided by RPL and can be used to specify the interactions between concurrent control subplans.

The control structures differ in how they synchronize subplans and how they deal with failures.

| Control Structure | Example of its usage |
|---|---|
| *in parallel do* $p_1 \ldots p_n$ | *in parallel do* NAVIGATE(⟨235,468⟩) BUILD-GRID-MAP() |
| *try in parallel* $p_1 \ldots p_n$ | *try in parallel* DETECT-DOOR-WITH-LASER() DETECT-DOOR-WITH-CAMERA() |
| *with constraining plan* p b | *with constraining plan* RELOCALIZE-IF-NEC() DELIVER-MAIL() |
| *plan* *with name* $N_1$ $P_1$ ... *with name* $N_n$ $P_n$ *order* $n_i < n_j$ | *plan* *with name* $S_1$ PUT-ONTABLE(C) *with name* $S_2$ PUT-ON(A,B) *with name* $S_3$ PUT-ON(B,C) *order* $S_1 \prec S_3$, $S_3 \prec S_2$ |

**Fig. 2.14.** Some RPL control structures and their usage

The *in parallel do* -construct executes a set of subplans in parallel. The subplans are activated immediately after the *in parallel do* plan is started. The *in parallel do* plan succeeds when all of his subplans have succeeded. It fails after one of its subplans has failed. An example use of *in parallel do* is mapping the robot's environment by navigating through the environment and recording the range sensor data. The second construct, *try in parallel* , can be used to run alternative methods in parallel. The compound statement succeeds if one of the subplans succeeds. Upon success, the remaining subplans are terminated. *with constraining plan* $P$ $B$, the third control structure, means "execute the primary activity $B$ such that the execution satisfies the constraining plan $P$." Policies are concurrent subplans that run while the primary activity is active and interrupt the primary if necessary. Finally, the *plan* -statement has the form *plan* STEPS CONSTRAINTS, where CONSTRAINTS have the form *order* $S_1 \prec S_2$. STEPS are executed in parallel except when they are constrained otherwise. Additional concepts for the synchronization of concurrent subplans include semaphores and priorities.

### 2.3.3 The Structured Reactive Controller

The Structured Reactive Controller (SRC), which is shown in Figure 2.15, is the top-level controller at the task-oriented layer. The main components of a structured reactive controller are the *process modules*, the *fluents*, the *structured reactive plan*, and the RPL *runtime system*. The elementary program units in the structured reactive controller are the process modules. Fluents, which are updated continually according to the messages sent by the modules of the perceptual subsystem and received by the SRC. The messages provide information about the current state of the other modules of the control system. The structured reactive plan specifies how the robot is to respond to events and feedback from the process modules to accomplish its jobs. The SRC is a collection of concurrent control routines that specify routine activities and can adapt themselves to non-standard situations by executing planned responses (Beetz and McDermott, 1994).

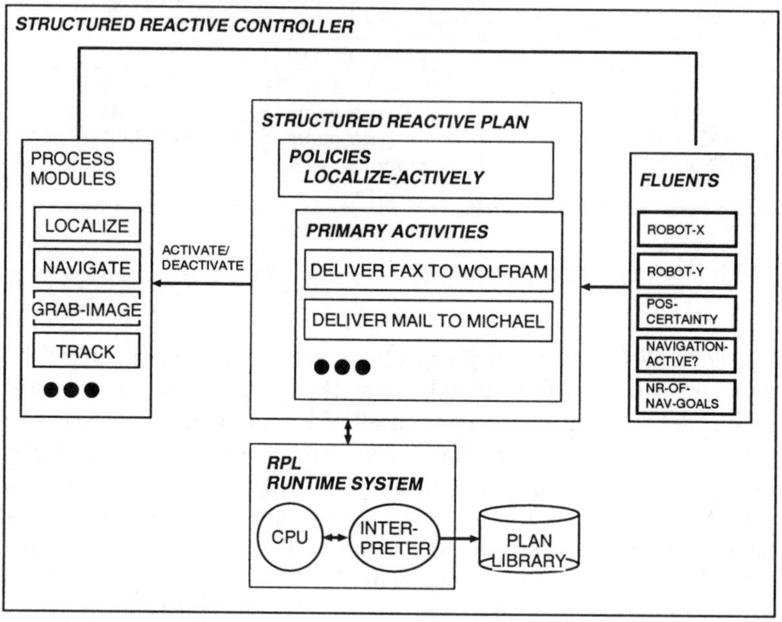


**Fig. 2.15.** Components of a structured reactive controller. The structured reactive plan specifies how the robot responds to changes of its fluents. The interpretation of the structured reactive plan results in the activation, parameterization, and deactivation of process modules

Structured reactive controllers work as follows. When given a set of requests, the structured reactive controller retrieves routine plans for individual requests and executes the plans concurrently. These routine plans are general and flexible — they work for standard situations and when executed concurrently with other routine plans. Routine plans can cope well with partly unknown and changing environments, run concurrently, handle interrupts, and control robots without assistance over extended periods. For standard situations, the execution of these routine plans causes the robot to exhibit an appropriate behavior in achieving their purpose. While it executes routine plans, the robot controller also tries to determine whether its routines might interfere with each other and watches out for non-standard situations. If it encounters a non-standard situation it will try to anticipate and forestall behavior flaws by predicting how its routine plans might work in the non-standard situation and, if necessary, revising its routines to make them robust for this kind of situation. Finally, it integrates the proposed revisions smoothly into its ongoing course of actions.

After having explained how the SRC manages the plan according to the changes of the robot's belief state, we will now describe how the plan gets interpreted. The interpreter executes the structured reactive plan, causing the process modules to be activated and deactivated. Threads of control become blocked when they have to wait for certain conditions to become true. For example, if the robot is asked to go to a location $x$ and pick up an object, the controller activates the behavior of moving towards $x$. The interpretation of the subsequent steps is blocked until the robot has arrived at $x$ (that is until the move behavior signals its completion).

Finally, we will look at how structured reactive plans (SRPs) is organized. SRPs are structured in a modular and transparent way so that automatic plan transformation techniques can easily retrieve subplans with particular roles — such as monitors or opportunities — and modify these plans without making them opaque for subsequent plan revision processes.

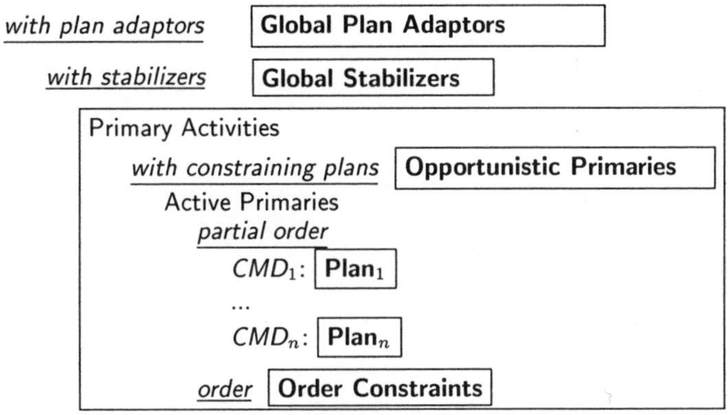


**Fig. 2.16.** The figure shows the block structure of an SRC. The activities that are needed to satisfy the user requests are contained in the active primaries. The execution of the active primaries is constrained by the opportunistic and the global policies

The RPL implementation of an SRC (Figure 2.16) consists of the ***primary activities*** that contain the plans for accomplishing the jobs and the surrounding ***global plan adaptors*** and ***stabilizers***. The primary activities are separated into the ***opportunistic primaries*** and the ***active primaries***. The ***active primaries*** are the ones that the robot is able to accomplish without help. The (partial) order in which the navigation subplans of the active primaries are executed is given by the ***order constraints***. The ***opportunistic primaries*** are the ones that the robot cannot accomplish by itself. Rather it has to wait for enabling conditions. For example, since RHINO cannot open doors it might have to wait for the opportunity of certain doors becoming open to complete its deliveries.

# 3. Plan Representation for Robotic Agents

In order to enable robot controllers to reason about and manipulate themselves, large fractions of the controllers have to be implemented as plans. Plans are symbolic specifications and descriptions of the robot's intended activity that can be reasoned about, revised, and executed. The robot uses plans as a resource for improving its problem-solving behavior. In this chapter, we will discuss the representation of plans that are intended to control autonomous robotic agents.

A plan representation consists of a plan language, a mechanism for interpreting plans written in this language, a set of inference mechanisms, and mechanisms for manipulating plans. The plan language provides the means for specifying the behavior of the robot. The inference mechanisms enable the planning mechanisms to infer information about a given plan. These inference mechanisms include ones for inferring goals from sub-plans and for predicting what will probably happen when the robot executes its plan. The plan manipulation mechanisms enable the robot to revise its intended courses of actions as its beliefs about the state of the environment change.

Plan representation for robotic agents is primarily concerned with two questions. First, what control patterns are necessary and sufficient for a robotic agent to solve a particular class of control problems? Second, how can the control patterns be represented to allow for effective plan generation, management, and execution?[1]

For the following discussion, we consider plan representation as a specific form of knowledge representation. Like domain knowledge representations, plan representations should be *representationally* and *inferentially adequate* and *inferentially* and *acquisitionally efficient* (Rich and Knight, 1991). We consider the representational adequacy of plan representations to be their ability to specify the necessary control patterns. Inferential adequacy is the systems' ability to infer information necessary for plan management and to perform the relevant plan management operations. Inferential efficiency is concerned with the time resources that are required to perform the plan management operations. Finally, acquisitional efficiency is the ability of plan representation systems to acquire new plan schemata and planning knowl-

[1] Steels (1984) has detailed a similar view for knowledge representation systems.

edge automatically. In this chapter, we concentrate on the design of the plan languages.

To be more specific, we believe that our requirement that plans should be able to improve the robot's problem-solving behavior imposes the following constraints on the design of plans.

1. ***Plans must be capable of controlling concurrent, interacting processes that have rich temporal structure, interact with the environment, and be event-driven.*** To make plans tolerant of sensor error, execution failures, and changing environments, plans can be implemented as collections of concurrent, event-driven control routines and specify how the robot is to respond to asynchronously arriving sensory data and other events. Many control patterns others than those provided by common plan representations have been proven to be necessary for flexible and reliable robot control (Firby, 1989).
   In addition to being capable of producing flexible and reliable behavior, the syntactic structure of plans should mirror the control patterns that cause the robot's behavior — they should be realistic models of how the robot achieves its intentions. Plans cannot abstract away from the fact that they generate concurrent, event-driven control processes without the robot losing the capability to predict and forestall many kinds of plan execution failures.
2. ***Plans should provide mechanism-specific control patterns.*** A successful plan representation for robotic agents must also support the control and proper use of the robot's different mechanisms for perception, deliberation, action, and communication. The full exploitation of the robot's different mechanisms requires mechanism-specific control patterns. Control patterns that allow for effective image processing differ from those needed for flexible communication, which in turn differ from those that enable reliable and fast navigation.
   To fully exploit the robot's different mechanisms, their control must be transparently and explicitly represented as part of the robot's plans. The explicit representation of mechanism control enables the robot to apply the same kinds of planning and learning techniques to all mechanisms and their interaction.
3. ***Plans must support economic inference and plan management.*** The generation of effective goal-directed behavior in settings where the robots lack perfect knowledge about the environment and the outcomes of actions and environments are complex and dynamic, requires robots to maintain appropriate plans during their activity. They cannot afford to entirely replan their intended course of action every time their beliefs change. Maintaining an appropriate and working plan requires the robot to perform various kinds of plan management operations including plan generation, plan elaboration, commitment management, environment monitoring, model- and diagnosis-based plan repair, plan failure

prediction, learning flexible and reliable plan schemata, a requirement that is stressed by Pollack and Horty (1999).
Unfortunately, most plan representations so far have been tailored for plan generation and have therefore been primarily designed to structure the search space that plan generation systems span. In contrast to these common plan representations, plan representations for robotic agents should enable the robotic agents to perform the plan management operations listed above both effectively and reliably.
The support of economic plan management is comprised of different aspects. First, the design of plans such that inferences that are common in plan management can be performed by pattern-directed retrieval rather than through the time consuming construction of long inference chains. Such inference tasks that are typical for plan management include the inference of the purpose of a given sub-plan, a sub-plan that is intended to achieve a given goal, and a sub-plan that causes some given effects. Second, the plans should enable the application of anytime plan management operations, which propose initial plan revisions quickly and tend to propose better revisions given more computation time (Zilberstein, 1996; 1995). The effective use of such anytime plan management operations requires plans to be prepared so that plan revisions can be proposed asynchronously while the plan is executed (Beetz and McDermott, 1996).

To summarize, we aim at developing a plan language that provides a uniform framework for controlling the different mechanisms of the robot and that is, at the same time, tailored for fast and reliable plan management. The plan language provides a uniform framework for mechanism control views routines for controlling the mechanisms as *actions* that need to be controlled, monitored, and synchronized and *plans* that can be generated, reasoned about, and revised. These features are crucial constituents of effective, fast, and robust mechanism control. Execution time plan management is supported representing sub-plans transparently and modularly, which enables pattern-directed access to plan pieces. In addition, plans are designed to allow for revisions during their execution. Our design choice is to reduce the general-purpose emphasis of plan languages for the sake of performing fast inferences on plans written in expressive plan languages.

In order to achieve the desiderata that we have outlined above, we use, extend, and specialize the XFRM plan representation (McDermott, 1991; McDermott, 1992b; Beetz, 2000). The extended XFRM plan representation provides the following components. First, it provides an interface to low-level control system based on the concept abstractions for *control processes*, *state variables*, and *events*. The plan language offers many of the control abstractions provided by modern event-driven programming languages including loops, program variables, processes, and subroutines as well as high-level constructs (interrupts, monitors) for synchronizing parallel actions. A *macro mechanism* allows for abstracting away from common and complex control patterns by means of defining sub-plan macros. Finally, the system provides a mechanism for implementing *inference procedures* and *plan revision rules*; and a framework for integrating *plan management operations* into the plan language.

Using the XFRM plan representation, we will develop in this chapter a plan language for a robotic agent by first, defining the interface between the plan representation and the robot's different mechanisms by specifying the control processes that the mechanisms provide, the state variables that characterize the state of a mechanism, and the events that the mechanism generates. Second, we develop control patterns for the flexible and reliable control of mechanisms. The control patterns for flexible and reliable execution that often yield complex sub-plans are then transformed into modular and transparent sub-plan schemata by using *sub-plan macros* that restrict and modularly represent mechanism specific and application specific interactions between sub-plans. Using the sub-plan macros plans can be designed to allow for fast and transparent access to the relevant plan pieces and successful continuations after revisions and interruptions. Finally, we specify plan adaptors for plan management and integrate them into the overall plan.

In the remainder of this chapter we proceed as follows. The next section describes the interface of our plan representation with the different mechanisms provided by the RHINO control system. The plan representation itself distinguishes between two layers of plans: low-level plans and high-level plans. Low-level plans, which are described and discussed in section 3.2, are situation-specific plans that perform some extended activity, such as navigating to a given destination or loading a particular object. The high-level plans, which are explained in section 3.3 specify behavior that is effective under a wider range of circumstances and represent the behavior modularly and transparently to facilitate inferences about plans and their modification.

## 3.1 Low-Level Integration of Mechanisms

The lack of appropriate control abstractions provided by common plan representations brought researchers to side-step plan representations when they specified the control and integration of mechanisms other than those concerning physical robot actions. Complex behavior specifications have been represented and specified in other formalisms, even if planning methods were applied to the control of these mechanisms. As a consequence, the mechanisms became black boxes from the point of view of the high-level robot control system. Examples of such approaches are: the situation-specific configuration of image processing routines (Firby et al., 1995); the use of dialogue grammars (Cole et al., 1997) and communication rules (Barbuceanu and Fox, 1995) that are compiled into finite state automata instead of using planned speech-act sequences (Cohen and Perrault, 1979); and the sequencing mechanism in layered hybrid architectures (Bonasso et al., 1997).

In this section, we will describe and discuss the nuts and bolts of integrating navigation, plan management mechanisms, communication, and image processing into the plan-based robot controller. The interfaces to the different mechanisms constitute an abstract robot control machine, which provides mechanism-specific fluents and process modules.

### 3.1.1 Navigation

One of the most crucial and essential capabilities of autonomous mobile robots is their capability to navigate reliably and effectively to specified locations within their environments. Navigation is the computation and execution of paths from the robot's current position to specified destinations. Detailed overviews on different approaches to robot navigation can be found in (Latombe, 1991) and (Borenstein, Everett, and Feng, 1996). The effective control of navigation with plans requires means for estimating the robot's position, computing and following paths, and for avoiding collisions.

We have seen in section 2.3.1 that the robot maintains a position estimate of its position in the environment. The estimate is represented as a probability distribution over the possible positions of the robot. For the purpose of high-level control, the probability distribution is abstracted into three state variables: the robot's position and orientation, the accuracy of the robot's position estimate (the diameter of the global maximum of the probability distribution), and the ambiguity (the number of local maxima that exceed a certain probability threshold). These state variables are stored in the fluents *ROBOT-X*, *ROBOT-Y*, *POS-ACCURACY*, and *POS-AMBIGUITY*.

The process module *ROBOT-GOTO* can be activated with a final destination and a sequence of intermediate destinations as its arguments. The sequence of intermediate destinations specifies constraints on the path that the robot has to take in order to reach the final destination. The navigation process performed by the process module causes the robot to move to the target points in the specified order. The module detects the successful completion of a navigation task by detecting that the robot is close enough to its final destination. A failure of the navigation task is signaled if the robot detects that it is unable to reach its destination (Firby, 1992; Firby, 1994). While active, the process module for navigation updates the fluent *NR-OF-REMAINING-DESTS* containing the number of remaining target points to be reached by the robot as a measure of the progress of the navigation process.

There are two more process modules that are provided by the low-level interface to the navigation system. The first one is the process module *POSITION-TRACKING* which continually estimates the robot's position and updates the fluents *ROBOT-X*, *ROBOT-Y*, *POS-ACCURACY*, and *POS-AMBIGUITY*. The sensing process is passive, which means that it does not generate any control inputs for the controlled process (Burgard, Fox, and Hennig, 1997). It only observes the odometry and other sensor readings and updates the position estimate accordingly. The position tracking process does not terminate unless it is deactivated by the controlling plan.

| **Fluents** | |
|---|---|
| *ROBOT-X*, *ROBOT-Y* | Fluents that store the values of the state variables for the robot's estimated x- and y-coordinates and the robot's orientation. |
| POS-ACCURACY | Accuracy of the position estimate in the global maximum of the probability distribution for the robot's position. |
| POS-AMBIGUITY | Number of local maxima with a probability that exceeds a given threshold probability. |
| REMAINING-PATH | Fluent that is updated by navigation processes. It contains the number of destinations that the robot still has to reach. |
| **Process Modules** | |
| ROBOT-GOTO | Process module that encapsulates navigation processes. |
| ACTIVE-LOCALIZATION | Module that directs the robot to explore locations in order to minimize the uncertainty in the robot's position estimate. |
| POSITION-TRACKING | Process module that passively tracks the robot's position estimate. |
| **Set Parameters** | |
| SET-TRAVEL-MODE | Command to reset the parameters of navigation processes. By setting the travel mode, the robot can be asked to drive very carefully, quickly, etc. |

**Fig. 3.1.** The integration of navigation mechanisms into SRCs. The table shows the fluents and process modules for interfacing the robot's navigation mechanisms. The fluents provide the robot with a concise representation of the current position estimate and the status of navigation tasks. The mechanism-specific process modules are the ones for active and passive localization and for navigation

The last process module is called *ACTIVE-LOCALIZATION* and its internal operation is described in (Beetz et al., 1998). The purpose of this process module is to resolve ambiguities and eliminate inaccuracies in the robot's position estimate. Given a probability distribution over the robot's current position that contains ambiguities and inaccuracies, the active localization process iteratively performs relative moves in order to maximally disambiguate the position estimate. The active localization processes causes, as a side effect, changes in the robot's position.

Finally, the interface to the navigation system provides a command for setting the mode of the navigation system. This mode is determined by two factors. The first factor is the choice of the sensors used for the detection of obstacles in the environment. The second component is some determination of how boldly or carefully the robot moves. It is often appropriate to move more cautiously in cluttered surroundings and more boldly in open areas.

The programming interface provided by the low-level integration of the navigation mechanisms is summarized in Figure 3.1.

### 3.1.2 Communication Mechanisms

RHINO's communication mechanisms comprise a restricted natural language facility which allows for the sending and receiving of electronic mails, means for transforming natural language text into sound files, displaying text and graphics on the screen of the onboard computer, and a button interaction facility which enables the robot to let buttons blink and monitor the buttons for being pressed.

The ability to perform natural language conversation significantly enhances robots' capabilities, augmenting both their abilities to perceive their environment and to effect changes within it. For instance, using natural language a robot can ask a person to open a door which it cannot open by itself. Being able to perform natural language communication also enables robots to receive a wider range of job specifications and acquire information that cannot be sensed using their sensors.

| **Fluents** | |
|---|---|
| LAST-EMAIL | The fluent contains the last electronic mail sent to the robot. |
| NEW-EMAIL? | The fluent is pulsed whenever the robot receives a new electronic mail. |
| **Process Modules** | |
| INTERPRET-SPEECHACT | The process module is activated with an email as its argument and translates the email into the internal speechact representation. |
| EXECUTE-SPEECHACT | The process module is activated with an internal speechact representation, translates the speechact into an email, and sends the email off to execute the speechact. |

**Fig. 3.2.** The integration of conversation mechanisms based on electronic mail. The interface contains two fluents: one is pulsed whenever a new electronic mail for the robot has arrived and the other one stores the last electronic mail. The process modules for controlling conversation contain one module for interpreting emails and translating them into an internal speechact representation and another one for executing a spechact by sending a corresponding email

We have chosen electronic mail as the primary channel to communicate with people in its environment. Compared to spoken language, electronic mail allows for some important simplifications of the natural language communication task. The email is already a sequence of words. It is more reasonable to ask for writing correct English sentences. The identification of the dialog partner is simple: the name is contained in the *sender* line of the mail header. Finally, the real-time requirement for electronic mail is less challenging than in spoken language dialog.

The electronic mail communication mechanism (Beetz and Peters, 1998) is integrated through the fluents *LAST-EMAIL*, which contains the last electronic mail sent to the robot and *NEW-EMAIL?*, which is pulsed whenever the robot receives a new email. The extension further contains the two process modules *INTERPRET-SPEECH-ACT* and *EXECUTE-SPEECH-ACT*. *INTERPRET-SPEECH-ACT* takes an electronic mail and translates it into a speechact with the content transformed into the robot's internal representation. *EXECUTE-SPEECHACT* takes an internally represented speechact, transforms it into an email, and sends it off.

**Interpreting Electronic Mails.** The interpretation of electronic mails proceeds in three steps: (1) transforming the electronic mail into an internal speech act representation; (2) parsing the content of the speech act; and (3) computing and analyzing the meaning of the speech act using the SATD map (see section 2.2).

**From:** peters2@cs.uni-bonn.de
**Date:** Fri, 24 Oct 1997 12:03:57
**To:** rhino+tcx@cs.uni-bonn.de
**Subject:** Library book

Could you please bring the yellow book on the desk in room a-120 to the library before 12:30

(a)

```
request
  :sender (THE PERSON
             (FIRST-NAME Hanno)
             (LAST-NAME Peters))
  :receiver (THE ROBOT
             (NAME RHINO))
  :time (THE TIME-INSTANT
             (DAY 24) (MONTH OCT)
             (TIME 12:03))
  :reply-with "Re: Library Book"
  :content content of the email
  :deadline (A TIME-INSTANT
             (BEFORE
               (TIME 12.30))))
```

(b)

**Fig. 3.3.** A typical electronic mail sent to the robot (sub-figure (a)) and its internal speech act representation (sub-figure (b)). The speech act representation indicates that the email contains a request and explicitly represents the sender, receiver of the email and the time of receipt. The representation of the email content is shown in figure 3.4

*Step 1: Transforming Electronic Mail into Internal Speech Acts.* Emails are represented internally as KQML statements. KQML (Knowledge Query and Manipulation Language) is a standardized language for the exchange of information between different knowledge-based systems (Finin, Labrou, and Mayfield, 1995). KQML messages communicate whether the content of a message is a request, order, etc (Cohen and Levesque, 1995). The robot can receive and perform requests, questions, acknowledgments, replies, and informative performatives.

The advantage of having representations of the purpose of electronic mails that abstract away from the contents of the mail is that this way the dialog behavior can be more concisely specified. In particular, at this level of abstraction we can specify which performatives are valid answers, whether responses are required, and so on.

Figure 3.3(left) shows a typical electronic mail sent to the robot. Its internal representation using KQML is shown in figure 3.3(right). The algorithm for constructing KQML statements from emails determines the type of speech act, extracts the sender, the subject, and the time when the email was sent, transforms the pieces of information into the corresponding KQML statement.

*Step 2: Parsing the Content of Electronic Mails.* After the KQML representation for an electronic mail is constructed, the content is parsed and transformed into a representation internal to the robot. This structured representation facilitates reasoning, retrieval, and plan construction. It provides logical connectives, is hierarchically structured and distinguishes between object descriptions, location descriptions, state descriptions, event specifications, and others.

```
(ACHIEVE
   (LOC (THE BOOK
              (COLOR YELLOW)
              (ON (THE DESK
                     (OWNER (THE PERSON
                                   (NAME MICHAEL)))
                     (IN (THE ROOM (NUMBER A-120))))))
        (THE LOC (IN (THE ROOM
                          (FUNCTION LIBRARY)))))
```

**Fig. 3.4.** Representation of the content of the electronic mail shown in Figure 3.3(a). It says that it is to be achieved that the object satisfying the description given as the first term of the LOC predicate will be at the location given by the second term

We assume that the content of the electronic mail is written in correct English that satisfies the following additional restrictions. The body of the email is a single, self-contained sentence. Only a subset of correct English sentence constructions is accepted. Furthermore, the vocabulary is restricted to correctly spelled words from the office delivery domain. The parser signals an error if the sentences are not parsable. These failure signals can then be caught by control structures of the plan that make the plan tolerant against such failures. Recovery strategies might then ask the person sending the electronic mail for clarifications using examples of parsable sentences.

The contents of electronic mails are parsed using a Definite Clause Grammar (Pereira and Warren, 1980). In particular, we use an extension of the grammar described in (Russell and Norvig, 1995) and (Norvig, 1992). The grammar is extended by augmenting the vocabulary relevant for office delivery tasks (most notably rules for parsing temporal statements, room names, etc.). The second extension of the grammar enables the robot to parse the sentences it generates (see next section). Besides extending the expressiveness of the language, we reduced the time required for grammatical processing using pattern matching techniques to filter out irrelevant grammar rules. Finally, we made the parser interruptible so that parsing processes can be terminated prematurely by deactivating them.

*Step 3: Interpreting the Content of Electronic Mails.* The next step in the interpretation of electronic mails is the identification of the objects the office courier has to pick up and the locations where they are to be picked up or to which they are to be delivered. To discuss how RHINO interprets an object description generated by the parsing step consider the following example:

```
(THE BOOK
     (COLOR YELLOW)
     (ON (THE DESK
              (OWNER (THE PERSON
                          (NAME MICHAEL)))
```

This object description contains a perceptual description of an object to be localized (LO), a description of a reference object (Michael's desk) and the spatial relation between them (cf. (Gapp, 1994; Stopp et al., 1994)). Since in spatial descriptions reference objects are usually stationary objects with salient features, RHINO requires reference objects to be contained in its environment model.

Thus, as a first step, the semantic interpretation component extracts the descriptions of the reference objects and retrieves in the second step the set of objects in the environment model that satisfy these descriptions. If the query returns more than one map object, the job description is ambiguous. If the query returns no map object then the job specification is inconsistent with the robot's model of the environment. In both cases RHINO has to start a dialog to clarify the job specification.

**Generating and Executing Speech Acts.** For the generation of speech acts we use an extension of the language generation facilities of DUCK (McDermott, 1985a). Essentially DUCK takes formulas written in predicate calculus and translates them into pseudo English sentences. To obtain better translations, the programmer can define skeletons for expressing predicates which are then compiled into rules. The rules for expressing logical connectors and handling the recursive structure of predicate calculus expressions are built into the system.

We have extended the facilities of the DUCK pseudo-English generator to enable it to handle object and location descriptions and the different kinds of speech acts that the RHINO control system can generate. Thus, a typical object description like

```
(THE BOOK (COLOR YELLOW)
          (ON (THE DESK (SIZE BIG)
                        (IN (THE ROOM (NUMBER A-120))))))
```

is translated into "the yellow book on the big desk in room A-120."

Since in electronic mail dialogs the content of an electronic mail is included in the answer, we have extended the parser so that it can parse the sentences that the pseudo-English generator produces.

### 3.1.3 Execution Time Planning

Execution time planning is integrated by providing two process modules (Beetz and McDermott, 1996): one for starting planning processes on sub-plans and one for installing revisions in a concurrently executed plan. Planning processes behave like sensing routines. Whenever a better plan is found, the planning process sends a signal. Planning processes provide and continually update their best results in the fluents *BEST-PLAN* and *BETTER-PLAN?*.

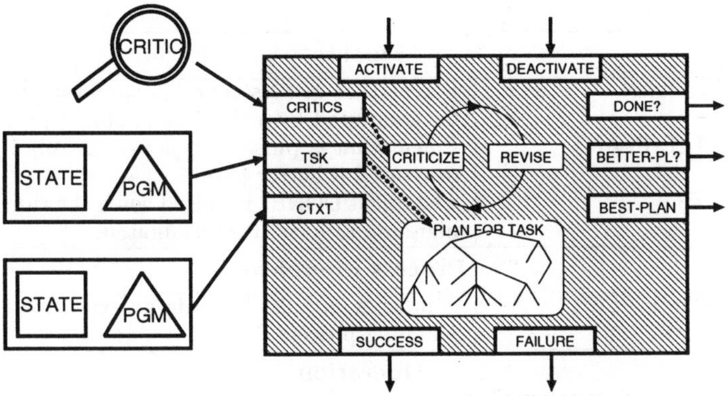


**Fig. 3.5.** Process module for the execution time planning of local sub-plans. The module is activated with the computational task for interpreting the sub-plan. The task data structure contains the plan as well as the data required to reconstruct the robot's belief state. The critics parameter specifies the kind of flaws the planning module should try to eliminate and the parameter CTXT specifies the context for which the sub-plan should be replanned

Figure 3.5 pictures the realization of execution time planning as a process module. The planning process can be started and deactivated. It takes the task for which a better plan is to be found and the critics to be used as arguments. The planning process automatically updates three fluents. The fluent *BETTER-PLAN?* is set true whenever the planner believes that it has found a better plan. The fluent *DONE?* is false as long as the planner is still looking for better plans. The fluent *BEST-PLAN* contains the best plan for the given task that the planner can offer at this moment.

The process module *RUNTIME-PLAN* enables the robotic agent to plan in a flexible and focused manner: it can reflect on how to perform its jobs while accomplishing them, focus on critical aspects of important subtasks, and ignore irrelevant aspects of their context. It can also postpone planning when lacking information, reconsider its course of action when noticing opportunities, risks, or execution failures, and integrate plan revisions smoothly into its ongoing activities.

The uniformity of the interfaces to navigation and execution time planning and the control structures provided by RPL enable a programmer to concisely specify a wide spectrum of interactions between planning and execution. RPL becomes a single high-level language that can handle both planning and execution actions. Providing the extensions for local planning of ongoing activities as primitives in RPL enables RPL's control structures to control not only the robot's physical actions but also its planning activities. It also allows RPL plans to synchronize threads of plan execution with local planning processes and to specify the communication between them.

| **Fluents** | |
|---|---|
| BEST-PLAN | Fluent that always contains the best plan so far computed by a runtime planning process. |
| BETTER-PLAN? | Fluent that is pulsed whenever the robot has generated a better plan. |
| DONE? | The fluent DONE? is pulsed when the runtime planning process has terminated. |
| **Process Modules** | |
| RUNTIME-PLAN | A process module that runs an execution time planning process. |
| **Operation** | |
| SWAP-PLAN | An operation that takes a plan and a subtask of the overall plan as its arguments and installs the plan for the execution of the subtask. |

**Fig. 3.6.** The integration of execution time planning into SRCs. The fluents represent the computational state of the planning process: the best plan it has produced so far, whether it has produced better plans, and whether the planning process is completed. The process module is called RUNTIME-PLAN. An additional operation SWAP-PLAN integrates new plans into the ongoing activity

### 3.1.4 Image Processing

To accomplish its jobs, the office delivery robot must also process various sorts of visual information. Consider, for instance, the following job: "get the red letter from Dieter's desk." Such a job specification contains visual descriptions of the objects to be delivered as well as the locations where these objects are to be picked up and delivered. Thus, the job could be accomplished by a plan of the following form:

1. **find** Dieter's desk and approach it
2. when you're close enough, **search** the desk **for** a red letter
3. if you can't find the book then
   - **look for** a person in the office and ask for help
   - if the person finds the book
     then **take** the book from the person

A brief analysis of the plan above shows that the office delivery robot has to perform different kinds of visual tasks like recognizing pieces of furniture and parts thereof, visually searching regions in images, detecting people (perhaps through motion detection), and tracking books in streams of camera images. Moreover, visual routines are closely coupled to the context of the control program. Visual routines must be synchronized with other actions of the robot. The control system has to recover locally from the failure of a visual routine to accomplish its job, or start additional visual routines to recover from execution failures. In addition, the robot often has to run several visual routines concurrently, which causes problems in the usage of restricted resources, such as access to cameras and processing time.

To realize plan-based control of image processing, we have extended RPL with an abstract machine that allows the programmers of the control software for autonomous robots to implement the image processing routines necessary for accomplishing a particular task themselves (Beetz et al., 1998).

The integration of image processing mechanisms is, in several aspects, different from the integration of other mechanisms. The main reasons for these differences are that image processing is a computationally very difficult problem and that the general interpretation of images captured by the camera is not feasible. Image processing by RHINO is therefore performed using libraries of special-purpose image processing routines that can be combined to perform more complex image processing tasks. Furthermore, unlike navigation tasks where the robot can only perform one navigation task at a time, an autonomous robot often performs several image processing tasks concurrently. There are also two kinds of image processing tasks. One type is that of image transformation tasks that receive an image captured by the camera and transform the image into other images and information. A second one is that of reactive image processing tasks that monitor sequences of images for the occurrence of certain events, such as gestures.

These characteristics of image processing for advanced service robot applications have consequences for their integration into structured reactive controllers. Let us consider the following program segment that grabs a *240 × 320* pixel image with the left camera, returning an image as an example. The segment activates the module *GRAB-IMAGE*, waits for the image capture to complete and returns the value of the fluent *IMAGE-ID-FL*. The program segment operates as follows. It creates two local fluents *IMG-ID-FL* and *DONE-FL*, which it passes as call parameters to the process module *GRAB-IMAGE*. The module sets *IMG-ID-FL* to the index of the grabbed image and pulses the fluent *DONE-FL* upon completion. Note, that image processing routines do not communicate via predefined fluents. Rather the control plans locally create the fluents for communication and synchronization.

```
with local fluents img-id-fl   ← NULL
                   done-fl     ← NULL
  do
    GRAB-IMAGE(:CAMERA  :LEFT
               :SIZE  240 × 320
               :COLOR  :TRUE
               :GRABBED-IMAGE  img-id-fl
               :DONE  done-fl)
    wait for done-fl
    FLUENT-VALUE(img-id-fl)
```

The interface provides operations for loading and unloading image processing functionality, process modules that perform primitive image processing operations and can be assembled into larger routines, and routines for vision-based monitoring.

The operations *LOAD-RECIPE-MODULE* and *UNLOAD-RECIPE-MODULE* load and unload a given module of image processing routines respectively.

Image processing operators that transform images and extract information from images are provided as process modules. The modules differ in the types of parameters they accept and the types of results they compute. The process module *GRAB-IMAGE*, for example, which we have used in the code segment discussed before, is activated with a camera name, image size, and whether the image is to be color or gray-scale as its parameters. The module updates the fluents *IMAGE*, in which the grabbed image is stored, and *DONE?*, which is pulsed upon process completion. Another process module, *PROCESS-IMAGE*, takes an image, a region of interest *ROI* to be processed, the image processing operator, and additional parameters for tuning as its arguments. The image processing operators include edge detectors (Sobel, Canny, ...), line followers, segmentation operators, and so on. The image and region of interest parameters are the indices of the images and regions of interest to be found in the image and ROI tables of the RECIPE module. The fluents that are updated by the *PROCESS-IMAGE* module are the *IMAGE*, *ROI*, and

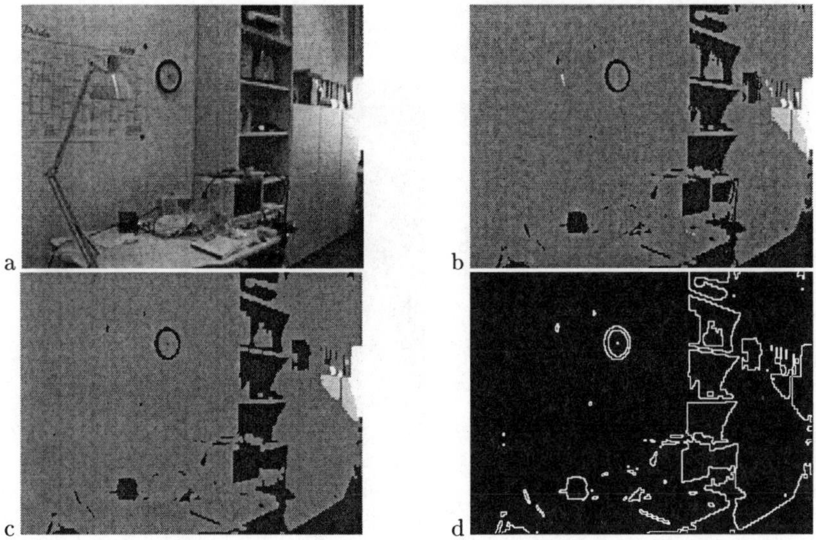

**Fig. 3.7.** Results of alternative image processing operation sequences. The operation sequences can be specified as sub-plans and performed in parallel. Concurrent reactive plans can then specify more effective image processing routines by selecting the sub-plans in situation specific ways and monitor the evidence for the results that is accumulated by the different sub-plans

*DONE* fluents that contain the image resulting from the application of the operator and the done signal. There are also process modules for matching lines, generating regions of interest, segmenting images, displaying images, and so on. There are also fluents that store the different kinds of image data such as images (or rather, indices for images), regions of interest, lines, and segments.

The process module *START-WATCHING-OUT-FOR-VISUAL-EVENTS* takes several parameters, one of which is a visual event description such as *mark-of-laser-pointer*, *active-computer-screen*, *image-region-with-motion*, *thumb-detected*, and *arm-pointing-to-side*. In addition to the event class description the operation requires a list of fluents that are used for synchronization and communication between the image processing process and the thread of control in which it is embedded. This interaction between image processing and the controlling process is realized by three fluents. The first one contains the status of the image processing process: whether the process is ready, active, interrupted, or terminated. The second fluent is pulsed whenever the image processing subsystem has detected a new visual event. Finally, the third fluent contains a description of the last visual event that has been detected. The data structure representing such descriptions is called a *designator*. The components of a designator comprise its local identifier used by the image processing process, the class of event (laser pointer mark, arm pointing gesa

b

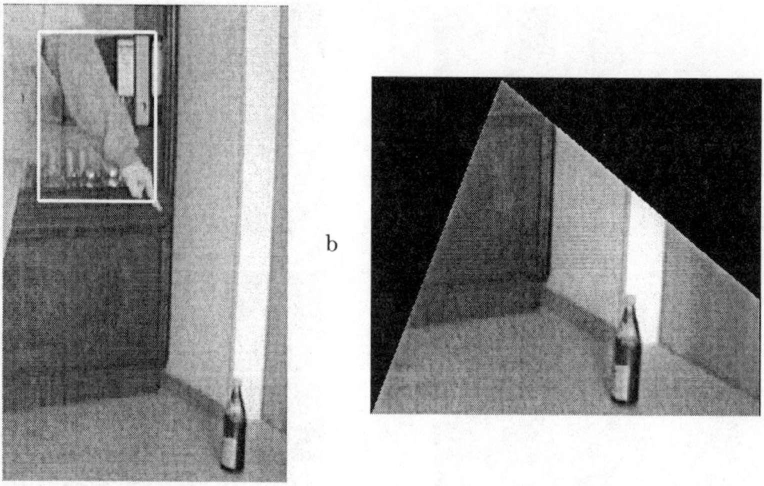

**Fig. 3.8.** Results produced by process modules that track pointing gestures. Pointing gestures (a) in image sequences produce target areas (b)

ture, and thumb gesture), and the bounding box of the event in the image. *STOP-WATCHING-OUT-FOR-VISUAL-EVENTS* deactivates a visual monitoring process.

### 3.1.5 Summary of Low-Level Integration

In this section we have described how the robot's mechanisms can be integrated into the robot plans. The integration consisted of specifying for each mechanism a set of process modules that constitute the interface to the control processes of this mechanism and a set of fluents that represent the state variables that are affected by the mechanism.

An integration of robot mechanisms through process modules and fluents is, in various ways, more general than an integration through predefined discrete actions. First, process modules can specify continuous control and be executed concurrently. Second, the fluents enable the plan-based controller to monitor and change the execution of active process modules. Because the process modules generate upon termination success and failure signals, they can be used as basic plan units and combined into more complex plans by the use of RPL control structures.

## 3.2 Low-Level Plans

In our approach to the plan-based control of robotic agents we distinguish between low-level and high-level plans. The high-level plans are designed to facilitate execution time plan management and use low-level plans to accomplish their purpose.

In this section we will describe and discuss the realization of low-level plans. Typically, low-level plans in SRCs are situation-specific plans for the direction of a control process. Low-level plans realize flexible and reliable control for mechanism- and application specific tasks. They monitor the execution of a control process and perform situation-specific reparameterization. They also perform trivial local failure recovery and signal failures if such a local recovery is not possible.

Let us consider a simple low-level plan for turning the robot to travel into the opposite direction as a simple example. Our plan makes use of a simple function →FACING-DIR? that, given a goal direction and an expected tolerance, returns a fluent that signals when the robot is facing the goal direction. Using this function, we can generate a rotation task specific fluent as we have done in the low-level plan TURN-BACKWARDS. The low-level plan TURN-BACKWARDS first starts the rotation. It then waits for the fluent to signal the completion of the rotation and then issues a command to stop the robot.

```
Function →FACING-DIR?   returns fluent   (GOAL-O, TOLERANCE)
  CREATE-FLUENT(TOLERANCE <| ROBOT-O − GOAL-O |)
```

```
RPL procedure TURN-BACKWARDS( )
  with local fluents turned-far-enough?
                       ←  →FACING-DIR?
                             (REMAINDER(ROBOT-O + 180, 360), 5)
     do
        ROBOT-START-TURN-RIGHT()
        wait for turned-far-enough?
        ROBOT-STOP
```

Unfortunately, this simple low-level plan is neither robust against interruptions nor flexible and failure tolerant. These properties can be achieved by wrapping the basic commands into the appropriate control structures.

```
RPL procedure TURN-BACKWARDS( )
with valve wheels
    do with local fluents turned-far-enough?
                        ← →FACING-DIR?
                            (REMAINDER.
                              (ROBOT-O + 180, 360),
                            5)
         do
             n-times 3
                do
                   ROBOT-START-TURN-RIGHT()
                   try in parallel
                      wait for turned-far-enough?
                      wait for robot-stopped?
                      wait time 10
                   ROBOT-STOP
               until turned-far-enough?
             if ¬turned-far-enough?
               then fail ":uncompleted-turn :angle ..."
```

In the remainder of this section we will describe and discuss the implementation of flexible and reliable low-level plans for different mechanisms.

### 3.2.1 Low-Level Navigation Plans

We will now look at the problem of specifying adaptive and constrained low-level navigation plans. To achieve execution time adaptability, we consider navigation behavior to be the result of interacting and concurrent control processes: the first one causes the robot to follow a path and the second one asks the robot to adapt its behavior as it passes certain regions.

Figure 3.9 pictures such a concurrent reactive navigation plan (Beetz et al., 1998). The plan consists of two components. The first one specifies a sequence of target points (the location indexed through the numbers 1 to 5 in Figure 3.9) to be reached by the robot. The navigation between the target points is accomplished by a standard path planner (Thrun et al., 1998). The second component specifies in detail when and how the robot is to adapt its travel modes as it follows the navigation path. In many indoor environments it is advantageous to adapt the driving strategy to the surroundings: to drive carefully (and therefore slowly) within offices because offices are cluttered; to drive with maximal clearance in doorways; and to drive quickly in the hallways. This part of the plan is depicted through regions with different textures for the different travel modes "office," "hallway," and "doorway." Whenever the robot crosses the boundaries between regions the appropriate travel mode is set in our collision avoidance module (Beetz et al., 1998).

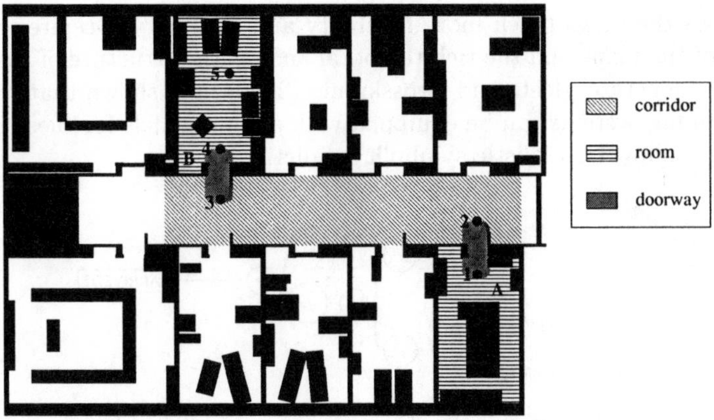


**Fig. 3.9.** Topological navigation plan for navigating from room A to B with regions indicating different travel modes and small black circles indicating additional navigation path constraints

The subsequent plan sketches the RPL code of a navigation process specification for leaving the office. The two components following the prescribed path and adapting the travel mode are implemented as concurrent sub-plans. The second component uses a fluent to measure the distance to the center of the doorway and two dependent fluents that signal that the robot enters and leaves the doorway. Initially, the travel mode is set to "office." Upon entering and leaving the doorway the travel mode is adapted.

```
concurrent reactive plan ( )
  in parallel do
      sequentially
          do START-NAV-PLAN(1 2)
              wait for (navigation-completed?)
      with local fluents distance-to-doorway
                          ← FLUENT-NETWORK(| ⟨x, y⟩ − ⟨x_dw, y_dw⟩ |)
          do with local fluents entering-dw? ← distance-to-door < 1m
                                entering-hw? ← distance-to-door > 1m
                  do SET-NAVIGATION-STRATEGY(office)
                     wait for entering-dw?
                     SET-NAVIGATION-STRATEGY(doorway)
                     wait for entering-hw?
                     SET-NAVIGATION-STRATEGY(hallway)
```

The distinctive feature of this plan representation is that behavior is specified through the interaction and interference of concurrent sensor-driven control processes as it is usually done in behavior-based robotics (Arkin, 1991).

This gives the plans much more flexibility and reliability. Despite the complexity of the plans and the rich temporal and spatial structure of the effects that they generate, Beetz and Grosskreutz (2000) have shown that robot action planning systems can be equipped with action models for these kinds of plans that allow for realistic symbolic predictions.

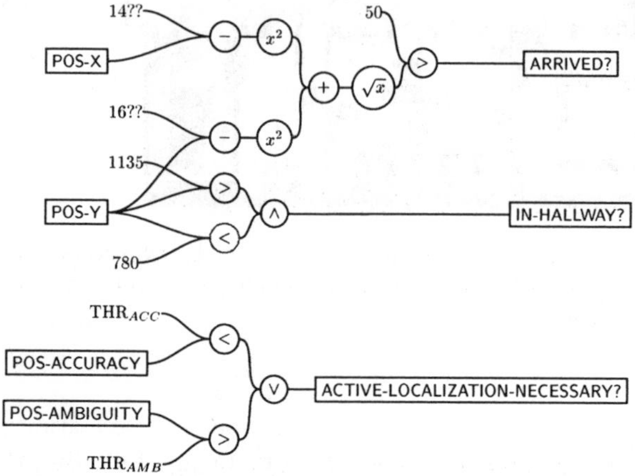


**Fig. 3.10.** A fluent network that computes qualitative task-specific percepts. The output fluents of the network compute whether the robot has already arrived at the destination of its current navigation task, whether it is in the hallway, and whether the position estimate is so inaccurate or unreliable that active localization is necessary

This raises a final issue. How does the robot detect qualitative events such as "passing a door," "a door being open," and "having arrived at the destination"? Such qualitative events are typically computed by fluent networks. Figure 3.10 shows fluent networks for the robot being in the hallway, having arrived at its destination, and having lost track of its position. The networks are essentially digital circuits through which the values of the fluents *ROBOT-X*, *ROBOT-Y*, *POS-ACCURACY*, and *POS-AMBIGUITY* are propagated, whenever their values change, in order to update the fluents that signal the occurrence of the above events.

### 3.2.2 Low-Level Image Processing Plans

To reason about image processing routines, they have to be implemented as plans. Image processing pipelines are common and appropriate means of representing complex image processing routines as plans because pipelines are modular, transparent and make the sequencing of image processing operations explicit (Rasure and Young, 1992). Image processing pipelines are data flow programs, directed graphs, in which each node represents an image processing operation and each directed arc represents a path over which data such as images, regions of images, or extracted lines, flow.

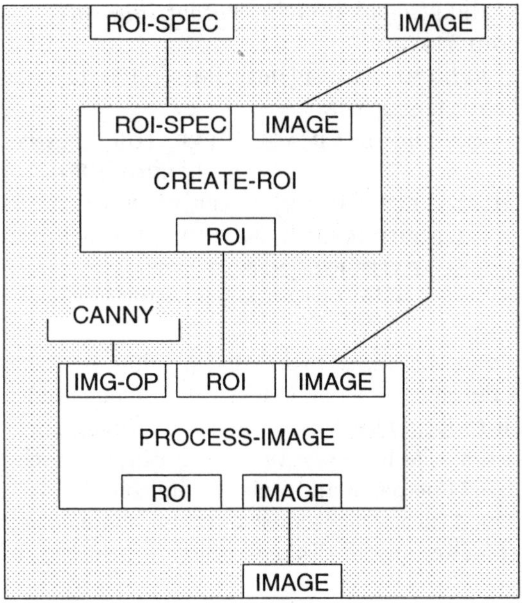


**Fig. 3.11.** A simple image processing pipeline containing two image processing steps: one for creating a region of interest in the image and a second one that applies a Canny edge detector to the selected region in the image

Figure 3.11 shows a simple image processing pipeline for detecting edges in a specified region of an image. The pipeline has to be provided with an image captured by the camera and a region of the image in which the edges are to be detected. These inputs to the pipeline have the names *ROI-SPEC* and *IMAGE*. The pipeline is a sequence of two operations. The first one, *CREATE-ROI*, extracts the region specified by *ROI-SPEC* from the image and provides the region named *ROI* as its outputs. The second pipeline step applies an edge detector (called Canny (Canny, 1986)) to the region. The image that is produced by this step is the result of the pipeline.

*Image Processing Pipelines as RPL Plans.* As low-level plans, image processing pipelines are specified using the macro *IMAGE-PROCESSING-PIPELINE*, which has the form

IP pipeline PIPELINE-NAME(*pipeline interface*)
1 *pipeline steps*
2 *:connections* *pipeline connections*

where *Pipeline Interface* specifies the input and output data paths, *Pipeline Steps* the image processing operators, and the *Pipeline Connections* are the data paths between the pipeline steps. The following code shows the low-level plan that corresponds to the pipeline depicted in Figure 3.11 .

```
IP pipeline IP-EXAMPLE(:input img-in, :output img-out)
        :step CREATE-ROI
              RPL-CREATE-ROI(:BORDERS ⟨⟨100, 10⟩, ⟨240, 240⟩⟩)
        :step PROCESS-IMAGE
              RPL-PROCESS-IMAGE(:OPERATOR :CANNY
                                :THRESHOLD 80)
    :connections (:INPUT :IMAGE) → (PROCESS-IMAGE :INPUT :IMAGE)
                 (:INPUT :IMAGE) → (CREATE-ROI :INPUT :IMAGE)
                 (CREATE-ROI :OUTPUT :ROI)
                     → (PROCESS-IMAGE :INPUT :IMAGE)
```

The expansion of the image processing pipeline on page 64 into an RPL plan is shown in figure 3.12. Connections are implemented as shared fluents declared in the <u>with local fluents</u> statement: one image processing operator writes the fluent that the next one takes as an input. Each plan step starts the image processing operation and waits for its completion. The connections are then transformed into <u>:order</u> constraints for the <u>plan</u> statement so that a pipeline step can run only if all steps that produce the step's inputs have been completed.

Image processing pipelines can be procedurally abstracted into RPL procedures, which can then be called as subroutines of the overall plan.

*Controlling Image Processing Routines.* The following piece of code shows the synchronization of image processing routines with the rest of the robot control system. We use a sub-plan FIND-A-PREDICTED-LINE that executes an image processing pipeline.

```
with valve wheels
   do with valve camera
        do try in parallel
             sequentially
                do FIND-A-PREDICTED-LINE(pred-line-fl,
                                         ...,
                                         done?)
                wait for done?
             wait for confidence-fl > 0.7
             wait time 5
```

To avoid destructive cross-process interferences, no other process should be able to make the robot leave its current location or redirect the camera. This is done using *valves*, semaphores that processes must own in order to

```
Macro Expansion IP-EXAMPLE(img-in, img-out)
1  with local fluents create-roi-done-fl
2                      ← CREATE-FLUENT(F1, FALSE)
3                      create-roi-roi-fl
4                      ← CREATE-FLUENT(F2, FALSE)
5                      process-image-done-fl
6                      ← CREATE-FLUENT(F3, FALSE)
7                      process-image-image-fl
8                      ← CREATE-FLUENT(F4, FALSE)
9                      process-image-roi-fl
10                     ← CREATE-FLUENT(F5, FALSE)
11 do
12    drive (img-out, process-image-image-fl)
13    partial order named subplan CREATE-ROI
14                  do RPL-CREATE-ROI(FLUENT-VALUE(img-in)),
15                                    100, 10, 240, 240,
16                                    create-roi-roi-fl,
17                                    create-roi-done-fl)
18                     wait for create-roi-done-fl
19                  named subplan PROCESS-IMAGE
20                  do RPL-TRANSFORM-IMAGE
21                     (FLUENT-VALUE(img-in)),
22                      FLUENT-VALUE(create-roi-roi-fl)),
23                      2, 80,
24                      process-image-image-fl,
25                      process-image-roi-fl,
26                      process-image-done-fl)
27                     wait for process-image-done-fl
28    :order CREATE-ROI ≺ PROCESS-IMAGE
```

**Fig. 3.12.** The expansion of the image processing pipeline on page 64 into an RPL plan. The communication between the pipeline steps is realized through the local fluents that are created in the declaration part of the with local fluents construct (line 1-10). The pipeline steps are defined in the lines 13-27. Each step waits for the completion of the corresponding image processing process. The sequencing of the pipeline is achieved through the order clause (line 28) of the partial order plan

issue particular kinds of commands. Thus, any process that causes the robot to move to another place must first request the valve *WHEELS*, move the robot upon receiving the valve, and then release the valve after the completion of the movement.

Another important aspect of controlling image processing routines is the assignment of computational resources to image processing tasks. In the example above we use the *TRY-ALL* statement to do this. The *TRY-ALL* statement succeeds if five seconds have passed, the confidence fluent exceeds the confidence threshold of *0.7*, or if the *FIND-A-PREDICTED-LINE* routine has completed. Thus, the code segment above runs the *FIND-A-PREDICTED-LINE* pipeline until there is enough evidence for a correctly recognized object and at most for five seconds. The valves *WHEELS* and *CAMERA* ensure that the robot and the camera are not moved.

*Application of Alternative Methods.* We can apply RPL control structures in order to combine alternative IP methods, for example, different methods for recognizing a piece of furniture. If, for instance, the item has a distinctive color, the robot can first extract similarly colored segments and then apply the other operations only to these segments. The different methods can be applied in parallel or sequentially, until one method acquires enough evidence or the accumulated evidence of all methods surpasses a given threshold. This can be specified using such RPL control structures as *TRY-ALL*.

*Failure Recovery.* Robust control systems must deal with failures of their image processing routines. Consider the routine *FIND-A-PREDICTED-LINE* which will fail if the furniture has been moved, if the robot is uncertain about its position, if the piece of furniture is (partly) occluded, and so on. RHINO 's plans that execute the image processing routines try to detect if the routines fail and signal failures in these situations. Thus *FIND-A-PREDICTED-LINE* signals a failure if the predicted line could not be recognized or could be recognized only with a low confidence. In the case of failure, robust control structures can then check the robot's confidence in its position estimate or use laser and sonar range finders to detect possible occlusions and recover from recognition failures.

### 3.2.3 Low-Level Conversational Plans

Achieving the intended effects of speech acts often requires complex dialogs. For example, courier jobs given to a robot in natural language are sometimes ambiguous and have to be clarified. To equip a robot with reliable communication capabilities it is therefore insufficient to simply provide the necessary natural language understanding and generation capabilities. It is also necessary that the robot properly controls and synchronizes the execution of its speech acts, monitors their effects, and reacts properly to the speech acts performed by the people in its working environment.

Such control patterns for conversations are specified as low-level plans. The example below shows part of RHINO's overall plan highlighting the plan steps for performing or processing speech acts. We see two global policies that ensure that incoming emails are processed immediately and ambiguous commands clarified. The other two conversational actions are part of the robot's primary activities, the ones intended to accomplish the robot's tasks.

```
with policy whenever new-email-arrived?
                   do process incoming email
   do with policy whenever ambiguous-cmd?
                          do clarify command
         do ...
                                 ask for help
              ...
                        inform people
              ...
```

The robot generates conversation acts when it perceives situations that it cannot handle on its own. For example, a navigation plan might contain a sub-plan to ask for help when the door by which the robot intends to leave is closed. Another example that shows the advantages of tight integration of communication mechanisms into the robot control system is taken from an autonomous tour guide application. When the robot explains an exhibit and its distance sensors show that people are standing around the robot it points with its camera to the exhibit instead forming a natural language description of the exhibit. In this case the integration allows for the synchronized use of different modalities for communication purposes.

### 3.2.4 Task-Specific Low-Level Plans

Besides mechanism-specific low-level plans, RHINO also makes use of task-specific ones. The low-level plans for loading and unloading letters are typical examples. Basically, loading a letter works as follows. The robot stands at a desk and asks the person sitting at the desk to put the letter into the robot's container. The person then takes a letter and puts it into the robot's container. After loading the letter the person presses one of the buttons of the robot. Pressing the button signals the robot that the loading is completed and the color of the letter that is loaded. Loading the letter is decomposed into a basic plan *try-load-letter* that handles the interaction part and a second plan *load-letter* that calls *try-load-letter* and does the synchronization, failure handling, and the updating of the load status of the container and the object description of the letter.

```
RPL procedure TRY-LOAD-LETTER(LETTER-DESIG)
  n-times 2
      do
         EXECUTE-SPEECH-ACT
            (REQUEST :SENDER  ROBOT
                     :CONTENT (ACHIEVE (ON-BOARD LETTER-DESIG)),
            :CHANNEL :SOUND)
         ROBOT-ASK-BUTTON(params)
         try in parallel
            wait time 15wait for button-pressed?
     until button-pressed?
  if ¬button-pressed?
     then fail ":nobody-loaded-a-letter"
```

The low-level plan above repeatedly asks a person to load a particular letter described by *letter-desig* on to the robot until the letter is loaded but at most twice. The person is asked by executing a speech act of the type *request*. The letter is loaded if the person has pressed a button that indicates the color of the letter. The *load-letter* plan listed below makes sure that no other concurrent process moves the robot or interacts via the buttons while the letter is loaded. The first step in the plan is to call *try-load-letter*, the plan continues only if *try-load-letter* has been carried out successfully, otherwise it fails. If the loading was completed successfully, the location slot of the letter designator and the content of the robot's container is updated. The last step signals a failure if the robot is carrying two identical letters.

```
RPL procedure LOAD-LETTER(LETTER-DESIG)
  with valve wheels
     do with valve buttons
          do TRY-LOAD-LETTER(LETTER-DESIG)
             UPDATE-DESIG(LETTER-DESIG
                           :COLOR  pressed-button
                           :LOC  CONTAINER)
             UPDATE-ROBOT-CONTAINER(:ADD  letter-desig)
             if LENGTH(GET-DESIGS(:COLOR  pressed-button
                                     :LOC  CONTAINER))
                 > 1
               then
                   fail "carrying two identical letters"
```

### 3.2.5 Summary of Low-Level Plans

The implementation of different capabilities as sub-plans of SRCs has several advantages. The navigation behavior is made adaptive by using fluents that signal when the robot is entering an office, a doorway or the hallway and trigger the appropriate adaptations of the travel mode. The navigation behavior is made robust and interruptible by adding policies that detect interrupts and handle them. Concurrent navigation behaviors are synchronized through semaphores. The communication behavior is made robust by detecting situations which require communication, for instance when commands are found to be ambiguous. In these situations, the robot performs dialogues via electronic mail to clarify commands. During these dialogues, the robot has to wait until its dialogue partner answers and remind its partner if he or she forgets. The use of concurrency and event driven execution allow for the implementation of image processing routines as dataflow programs — a common method for the transparent specification of image processing strategies (Beetz et al., 1998). SRCs also allow for the situation-specific generation of special purpose routines, the coordination of the use of restricted resources—such as the cameras and computational resources—and the recovery from failed image processing routines.

In conclusion, SRCs use a single language (RPL) to control different capabilities and specify their internal operations. SRCs view these capabilities as actions that need to be controlled, monitored, and synchronized as well as plans that can be generated, reasoned about, and revised. These features are crucial constituents of the effective, efficient, and robust integration of different modalities of the robot.

## 3.3 Structured Reactive Plans

While low-level plans are suitable means for handcoding task-specific plans, they have been found to be complicated for autonomous learning and automatic plan management. Plan-based robotic agents need reliable and fast algorithms for the generation of mission-specific plans, the anticipation and diagnosis of execution failures, and the editing of sub-plans during their execution. In this section we develop modular and transparent representations that facilitate reasoning and plan management operations.

The robotic agent disposes of a plan library that contains plan schemata for achieving, perceiving, and maintaining some set of states *s*. For example, the plan library contains plans indexed by *achieve* LOC(ROBOT,⟨?X,?Y⟩), which asserts that the robot will be at location LOC(ROBOT,⟨?X,?Y⟩). If during operation the robot needs to go to a particular location, say LOC(ROBOT, ⟨100,100⟩), plan management operations can retrieve the plan schema, instantiate it, and insert the instantiated plan schema into the plan.

### 3.3.1 Properties of SRCs and Their Sub-plans

To avoid unsolvable computational problems, SRCs use methods that make strong assumptions about plans to simplify the computational problems. As a consequence, SRCs can apply reliable and fast algorithms for the construction and installment of sub-plans, the diagnosis of plan failures, and for editing sub-plans during their execution. Making assumptions about plans is attractive because planning algorithms construct and revise the plans and can thereby enforce that the assumptions hold.

In a nutshell, the set of plans that an SRC generates is the reflexive, transitive closure of the routine plans with respect to the application of plan revision rules. Thus, to enforce that all plans have a property $Q$ it is sufficient that the routine plans satisfy $Q$ and that the revision rules preserve $Q$.

We propose that SRCs use routine plans that are ***general***, ***embeddable***, ***transparent***, and ***interruptible***. These properties make it particularly easy to reason about the plans while the plans can still specify the same range of concurrent percept-driven behavior that RPL can. Below we will describe these properties and explain why they are important.

*1.* ***Generality.*** *A routine plan* **p** *with goal* **g** *is called* **general** *if it accomplishes its goal for* **typical situations** *and* **typical events** *that might occur while performing plan* **p**. Plans can be made general by equipping them with means for situation assessment, alternative situation-specific plans (Schoppers, 1987), and for failure detection and recovery (Firby, 1987). The routine plan $p$ for a goal $g$ should be implemented such that based on the percept history of the robot plan $p$ has the highest expected utility for accomplishing goal $g$ for the prior distribution of situations and events. At the moment the routine plans are handcrafted but in the long run they should be automatically learned. The use of general plans makes planning easier because general plans do not require the planners to accurately predict the contexts in which plan steps are to be taken.

*2.* ***Embeddability.*** *A routine plan* **p** *with goal* **g** *is called* **embeddable** *if it achieves it's goal* **g** *even when run concurrently with a set of routine plans* $p_1, \ldots, p_n$. Embeddability is important because, in general, concurrent control processes interfere and can thereby cause each other to fail. Thus, an embeddable plan for searching for a book on a table must prevent concurrent plans moving the robot or pointing the camera in different directions. Embeddability is typically achieved by adding synchronization mechanisms, such as semaphores, to the plans. Using embeddable plans the planner does not have to reason about all possible interferences among concurrent sub-plans. Note, however, that a set of concurrent embeddable plans does, in general, not guarantee that the conjunction of their goals holds after their successful conclusion.

*3.* ***Interruptibility.*** *A routine plan* **p** *with goal* **g** *is called* **interruptible** *if for any routine plan* **p'** *that does not make* **g** *unachievable, the plan* with policy whenever *f p' p accomplishes* **g**. To handle interruptions and their possible side effects, a routine plan $p$ must detect the interruptions, assess the situation after the interruptions have ended, and call the appropriate continuation to achieve its goal $g$. The use of interruptible sub-plans enables the planning algorithms to specify additional constraints for the execution of sub-plans in a modular way (using policies) without risking making sub-plans unexecutable.

*4. **Transparency.*** *A routine plan is called **transparent** if it has the form* reduce achieve $g$ $p$ *if and only if the purpose of a plan* **p** *is to achieve* $g$. Transparent plans support a planning algorithm in inferring whether the robot has tried to achieve, maintain, or perceive a given state. They specify the purpose of sub-plans in terms of states of the environment (using declarative RPL statements (Beetz and McDermott, 1992)) and enable planning modules to infer important aspects of the plan by syntactic matching and pattern-directed retrieval.

*5. **Locational Transparency.*** We call a plan ***location transparent*** if and only if every sub-plan $p$ that is to be performed at a particular location $l$ has a symbolic name and is reduced from a plan of the form at location $\langle x,y \rangle$ $p$. In addition, ordering conditions on at location sub-plans are either specified in the form of RPL statements or explicit :order clauses (see below).

For location transparent plans, a scheduler can traverse the plan recursively and collect all names of the at location sub-plans, the locations where they are to be performed, and the ordering constraints on these sub-plans. The scheduler then determines an extension on these ordering constraints that maximizes the expected utility of the overall plan and installs these ordering constraints in the plan.

### 3.3.2 High-Level Navigation Plans

Let us now investigate the representational issues in the development of high-level plans in the context of specifying navigation behavior. We will do so by stepwise developing a plan representation for an autonomous robot office courier. We will first make the low-level navigation plans more transparent and modular to support the autonomous acquisition of good environment specific navigation plans. In the next step, we will take these basic navigation plans and extend them so that they can be carried out opportunistically or interrupted by opportunistic plan steps. Finally, we will make the plans that employ these general and interruptible navigation routines more transparent and represent the navigation tasks in an entirely modular way.

**Structured Reactive Navigation Plans.** While concurrent reactive plans are suitable means for handcoding navigation plans, they have been found to be complicated for autonomously learning better navigation plans. We learned by experience that plan representations that support learning better navigation plans should contain modular sub-plans that correspond to the behavior stretches they control. They should also represent the parameters for the low-level navigation processes explicitly. Finally, in our case we wanted the plan representation to mirror the fact that a learned navigation plan is a default plan optimized for average performance with additional sensor-driven behavior modifications.

We call the plan representation that we have designed to satisfy these requirements ***structured reactive navigation plans*** (SRNP*s*). SRNPs specify a default navigation behavior and employ additional concurrent, percept-driven sub-plans that overwrite the default behavior while they are active. The activation and deactivation conditions of the sub-plans structure the continuous navigation behavior in a task-specific way. SRNPs have the following form.

*structured reactive navigation plan* *(s,d)*
    *with sub-plans* SUBPLAN-CALL(*args*)
        *parameterizations* $p_1 \leftarrow v_1$, ..., $p_n \leftarrow v_n$
        *path constraints* $\langle x_1, y_1 \rangle$, ..., $\langle x_m, y_m \rangle$
        *justification* *just*

    ...
  DEFAULT-GO-TO ( *d* ).

where $p_i \leftarrow v_i$ specifies a parameterization of the subsymbolic navigation system. In this expression $p$ is a parameter of the subsymbolic navigation system and $v$ is the value $p$ is to be set to. The parameterization is set when the robot starts executing the sub-plan and reset after the sub-plan's completion. The path constraints are sequences of points that specify constraints on the robot's path. The sub-plan call specifies when the sub-plan starts and when it terminates. To be more specific, consider the following SRNP.

*navigation plan* *(desk-1,desk-2)*
    *with subplans*
      TRAVERSE-NARROW-PASSAGE(⟨635, 1274⟩,⟨635, 1076⟩)
        *parameterizations* *sonar* ← *off*
            *colli-mode* ← *slow*
        *path constraints* ⟨635, 1274⟩,⟨635, 1076⟩
        *justification* *narrow-passage-bug-3*
      TRAVERSE-NARROW-PASSAGE(...)
      TRAVERSE-FREE-SPACE(...)
  DEFAULT-GO-TO ( *desk-2* )

The SRNP above contains three sub-plans: two for traversing doorways and one for speeding up the traversal of the hallway. A sub-plan for leaving an office is shown in more detail. The path constraints are added to the plan for causing the robot to traverse the narrow passage orthogonally with maximal clearance. The parameterizations of the navigation system specify that the robot is asked to drive slowly in the narrow passage and to only use laser sensors for obstacle avoidance to avoid the hallucination of obstacles due to sonar crosstalk.

The plan interpreter expands the declarative navigation plan macros into procedural reactive control routines. Roughly, the plan macro expansion does the following. First, it collects all path constraints from the sub-plans and passes them as intermediate goal points for the MDP navigation problem. Second, the macro expander constructs for the sub-plan call a monitoring process that signals the activation and termination of the sub-plan. Third, the following concurrent process is added to the navigation routine: wait for the activation of the sub-plan, set the parameterization as specified, wait for the termination of the sub-plan, and reset the parameterization.

**Interruptible and Embeddable Navigation Plans.** The navigation plans would be of little use if they could not be employed in diverse task contexts. This can be achieved by turning structured reactive navigation plans into interruptible and embeddable plans.

```
highlevel-plan achieve(loc(rhino, ⟨x,y⟩))
 1  with cleanup routine ABORT-NAVIGATION-PROCESS
 2      do with valve wheels
 3            do loop
 4                 try in parallel
 5                     wait for navigation-interrupted?
 6                     with local vars NAV-PLAN ← GENERATE-NAV-PLAN(c,d)
 7                         do swap-plan (NAV-PLAN,NAV-STEP)
 8                             named subplan NAV-STEP
 9                                 do DUMMY
10               until IS-CLOSE?(⟨x,y⟩)
```

The lines 6 to 8 make the navigation plan independent of its starting position and thereby more general: given a destination *d*, the plan piece computes a low-level navigation plan from the robot's current location *c* to *d* and executes it (Beetz and McDermott, 1996).

To run navigation plans in less constrained task contexts we must prevent other — concurrent — routines from directing the robot to different locations while the navigation plan is executed. We accomplish this by using semaphores or "valves," which can be requested and released. Any plan that asks the robot to move or stand still must request the valve *WHEELS*, perform its actions only after it has received *WHEELS*, and release *WHEELS* after it is done. This is accomplished by the statement *WITH-VALVE* in line 2.

In many cases processes with higher priorities must move the robot urgently. In this case, blocked valves are simply pre-emptied. To make our plan ***interruptible***, robust against such interrupts, the plan has to do two things. First, it has to detect when it gets interrupted and second, it has to handle such interrupts appropriately. This is done by a loop that generates and executes navigation plans for the navigation task until the robot is at its destination. We make the routine cognizant of interruptions by using the fluent *NAVIGATION-INTERRUPTED?*. Interrupts are handled by terminating the current iteration of the loop and starting the next iteration, in which a new navigation plan starting from the robot's new position is generated and executed. Thus, the lines 3-5 make the plan interruptable.

To make the navigation plan ***transparent*** we name the routine plan *ACHIEVE(LOC(RHINO, ⟨x,y⟩))* and thereby enable the planning system to syntactically infer the purpose of the sub-plan.

Interruptible and embeddable plans can be used in task contexts with higher priority concurrent sub-plans. For instance, a monitoring plan that our controller uses estimates the opening angles of doors whenever the robot passes one. Another monitoring plan localizes the robot actively whenever it has lost track of its position.

```
with policy whenever pos-uncertainty > 0.8
              do ACTIVATE(ACTIVE-LOCALIZATION)
                 wait for pos-uncertainty ≤ 0.8
  do [active-primaries]
```

**Delivery Tour Plans.** To facilitate online rescheduling we have modularized the plans with respect to the locations where sub-plans are to be executed using the *at location* plan schema. The [*at location* $\langle x,y \rangle$ $p$] plan schema specifies that plan $p$ is to be performed at location $\langle x, y \rangle$. Here is a simplified version of the plan schema for *at location* .

```
named subplan N_i
  do at location ⟨x,y⟩   p   by
     with valve Wheels
        do with local vars DONE? ← FALSE
              do loop
                    try in parallel
                       wait for Task-Interrupted?(N_i)
                       sequentially
                          do NAVIGATE-To⟨x,y⟩
                             p
                             DONE? ← TRUE
                 until DONE? = TRUE
```

The plan schema accomplishes the performance of plan $p$ at location $\langle x, y \rangle$ by navigating to the location $\langle x, y \rangle$, performing sub-plan $p$, and signaling that $p$ has been completed (the inner sequence). The *with valve* statement obtains the semaphore *Wheels* that must be owned by any process changing the location of the robot. The loop makes the execution of $p$ at $\langle x, y \rangle$ robust against interruptions from higher priority processes. Finally, the *named sub-plan* statement gives the sub-plan a symbolic name that can be used for addressing the sub-plan for scheduling purposes and in plan revisions. Using the *at location* plan schema, a plan for delivering an object $o$ from location $p$ to location $d$ can be roughly specified as a plan that carries out *pickup(o)* at location $p$ and *put-down(o)* at location $d$ with the additional constraint that *pickup(o)* is to be carried out before *putdown(o)*. If every sub-plan $p$ that is to be performed at a particular location $l$ has the form *at location* $\langle x,y \rangle$ $p$, then a scheduler can traverse the plan recursively and collect the *at location* sub-plans and install additional ordering constraints on these sub-plans to maximize the plan's expected utility.

To allow for smooth integration of revisions into ongoing scheduled activities we designed the plans such that each sub-plan keeps a record of its execution state and if started anew skips those parts of the plan that no longer have to be executed (Beetz and McDermott, 1996). We made the plans for single deliveries restartable by equipping the plan $p$ with a variable storing the execution state of $p$ that is used as a guard to determine whether or not a sub-plan is to be executed. The variable has three possible values: *to-be-acquired* denoting that the object must still be acquired; *loaded* denoting that the object is loaded; and *delivered* denoting that the delivery is completed.

```
if   ¬   EXECUTION-STATE(p, delivered)
  then sequentially
         do if EXECUTION-STATE(p, to-be-acquired)
              then AT-LOCATION   L   PICK-UP(o)
            if EXECUTION-STATE(p, loaded)
              then AT-LOCATION   D   PUT-DOWN(o)
```

A plan for a delivery tour has the form *plan steps orderings* where *steps* are instances of the delivery plan schema. *Constraints* have the form *:order* $s_1$ $s_2$ where the $s_i$s are name tags of sub-plans (e.g., the name tags of the *at location* sub-plans. *:order* $s_1$ $s_2$ means that $s_1$ must be completed before $s_2$ is started. *steps* are executed in parallel except when they are constrained to be otherwise.

We call a tour plan for a set of delivery commands restartable if for any sequence of activations and deactivations of the tour plan, terminating after the activation of the last plan step guarantees that (1) every pickup and putdown operation has been executed; (2) the order in which the sub-plans have been executed satisfies the ordering constraints; (3) every pickup and putdown operation has been executed at most once.

This kind of plan representation has been employed in the probabilistic prediction-based runtime rescheduling of robot courier tasks (Beetz, Bennewitz, and Grosskreutz, 1999). The representation allowed for extremely fast access to plan steps that need scheduling and the modularity of the representation allowed for a very simple installation of scheduling constraints. The fact that the navigation plans used by the *at location* sub-plan are embeddable and interruptible enabled a very flexible and reliable execution. The robot courier using the representation could be shown to outperform scheduling approaches that do no predictive planning.

### 3.3.3 Structured Reactive Plans for Other Mechanisms

**Structured Reactive Image Processing Plans.** Making their image processing capabilities efficient and reliable often requires robots to (1) use contextual information to simplify image processing tasks, (2) tailor image processing routines to specific situations, (3) tightly integrate the image processing capabilities into the operation of the control system, and (4) make optimal use of the robot's computational resources.

There are many image processing parameters that the robot can adjust by sensing but cannot predetermine. For example, the quality of lighting within a room can be perceived by the robot's sensors but the robot cannot know in advance whether a room will be adequately lit. Thus, the image processing operation sequences need to be tailored to fit the robot's particular circumstances and environment. Instead of requiring the image processing routines to have general applicability and reliability, RHINO applies a selection of context-specific image processing routines that work well in the contexts for which they have been tailored.

It is possible to dynamically modify the image processing pipelines because SRIPPs are plans and because specific tools provided by RPL permit processes to (1) project what might happen when a robot controller executes a SRIPP (McDermott, 1992b); (2) infer what might be wrong with a SRIPP given a projected execution scenario; and (3) perform complex revisions on SRIPPs (McDermott, 1992b; Beetz and McDermott, 1994)[2]. Thus, explicitly representing image processing routines in the robot's plan enables the robot to apply various methods to simplify its vision tasks. It can pose more specialized vision problems (Horswill, 1995) and adapt the computer vision strategies to the particular context (environment, task, image) during execution (Prokopowicz et al., 1996).

Image processing pipelines are also very similar to partial-order plans and causal link plans (McAllester and Rosenblitt, 1991) which are plan representations that are widely used by AI planning techniques. Techniques from these areas can therefore be applied to this tailoring process.

## 3.4 The Plan Adaptation Framework

This section presents an approach to runtime plan adaptation which is based on the following ideas:

1. Plan adaptation is performed by ***self-adapting plans*** that contain so-called ***plan adaptors***. Plan adaptors are triggered by specific ***belief changes***. Upon being triggered, the adaptors decide whether adaptations are necessary and, if so, perform them.

[2] These facilities have been fully implemented for the physical actions of the robot but are not yet fully applicable to the image processing routines.

2. Runtime plan adaptation is carried out as a two stage process that first makes a very fast ***tactical adaptation*** that decides on how to continue at that very moment, and then immediately starts a ***strategic adaptation process*** that may later (when it arrives at a decision) revise the tactical adaptation. The strategic decision reasons through the possible consequences of plan adaptations and performs revisions based on foresight.
3. Plan adaptation processes are specified explicitly, modularly, and transparently and are implemented using ***declarative plan transformation rules***.
4. The plans that the plan adaptation processes reason about and revise are ***adaptable***. They are designed to allow the plan adaptation processes fast and transparent access to the relevant plan pieces. Adaptable plans are also capable of continuing their execution after revisions.

The most salient advantages of our computational model of runtime plan adaptation are as follows.

- Plan adaptation processes are specified as sub-plans of the robot's overall plan. As a consequence, ***all*** decision making is represented explicitly within the plan and not hidden in a separate sequencing system, which enables the robot to add, delete, and revise in the same way as other plan steps. Furthermore, the interaction between plan adaptors and the plans is explicitly specified in the plan.
- Implementing plan adaptation as a combination of tactical and strategic adaptations also has advantages. The robot can exploit the fact that many delayed decisions, if they require revisions at all, can easily be incorporated into a plan because they only affect future parts of the plan that have yet to begin. The robot therefore does not have to stop to replan and is always under the control of a plan.

The remainder of this section describes and discusses the components of our plan adaptation framework, which is a rational reconstruction of the use of plan revision methods in SRCs (Beetz, 1999) and is also implemented using transformational planning of reactive behavior (McDermott, 1992b; Beetz and McDermott, 1994).

*Plan Adaptors.* The plan adaptation methods of a plan adaptor are divided into tactical adaptations that are performed instantaneously and strategic adaptations that may take a considerable amount of computation time to arrive at a decision. Thus, a plan adaptor is defined by

```
define plan adaptor      name    (args)
    :triggering-belief-change    bc
    :tactical-adaptation           t-adpt*
    :strategic-adaptation        s-adpt*
```

where *bc* is the triggering belief change, *t-adpt** a possibly empty set of tactical adaptation methods and *s-adpt** a possibly empty set of strategic adaptation methods.

Since plan adaptations might have global consequences that cannot be easily foreseen, tactical decisions are sometimes shortsighted. Therefore, plan adaptation is implemented as a two step process: first a tactical plan adaptation is performed which buys time and then a strategic adaptation is made that might overwrite the tactical decision. Of course, there are cases in which stopping the execution of the current plan and waiting for a new plan is the best execution strategy. This is just a special case of a plan adaptor, in which the tactical adaptation inserts an additional plan step into the plan that blocks further execution until the strategic adaptation methods have reached a decision on how to continue. To realize reactive decisions we specify plan adaptors without strategic adaptation methods.

*Adaptable Plans.* Our plan adaptation framework uses plan adaptation methods that make strong assumptions about plans in order to simplify the computational problems. As a consequence, plan adaptation methods can apply reliable and fast algorithms for the construction and installation of sub-plans, the diagnosis of plan failures, and for editing sub-plans during their execution. Making assumptions about plans is attractive because plan adaptation methods construct and revise the plans and can thereby enforce that the assumptions hold. To enforce that a self-adapting plan satisfies a property $Q$ throughout its existence, it is sufficient that it satisfies $Q$ at the beginning of execution and that its plan adaptors preserve $Q$ (Beetz, 1999). A robot office courier that schedules its activities in order to shorten the paths the robot has to follow, for example, might enforce that the locations at which actions have to be performed are represented explicitly and the actions to be performed at the same locations collected modularly in sub-plans. These properties facilitate runtime plan scheduling drastically.

*Self-Adapting Plans.* Self-adapting plans are specified using the construct

```
with plan adaptor    Adpt
    Pl
```

which means carry out plan *Pl* but whenever the triggering belief of the plan adaptation process *Adpt* changes *Adpt* is executed. For example, the following piece of plan

<u>*with plan adaptor*</u> *handle-unexpected-open-door({A-110,A-120})*
*delivery plan*

tells the robot to carry out *delivery plan* but if it learns that office A-110 or A-120 has become open it should then run the plan adaptation procedure *handle-unexpected-open-doors*.

Since self-adapting plans are themselves plans, other plan adaptors can reason about them, delete, insert, or revise them. For example, a global scheduler for office delivery jobs can add ordering constraints on the execution of the individual delivery steps but also on the plan adaptors for rescheduling that are triggered if the robot subsequently learns that assumptions underlying the schedule are violated.

The <u>*with plan adaptor*</u> sub-plan macro roughly expands into the pseudo code shown below. The expanded macro is a policy that tries to revise the SRP whenever the fluent *revision$_i$-necessary-trigger?* is pulsed. The result of the rule application is stored in the variable *PROPOSED-REVISION*. If a revision is proposed, the fluent *new-srp?* is pulsed. To allow for concurrent revision the SRP is embedded in an endless loop. An iteration is completed if the SRP is completely executed or a plan revision proposed. Each iteration first swaps the proposed plan into the SRP and executes it afterwards. This plan swapping in the midst of execution can be performed because the SRP is revisable.

```
whenever revision_i-necessary-trigger?
   do with valve revision-authority
          do with local vars PROPOSED-REVISION
                                      ← APPLY(rev-rule_i,srp)
                 do if PROPOSED-REVISION
                        then NEW-SRP ← PROPOSED-REVISION
                             pulse (new-srp?)
```

*Plan Adaptation Rules.* The plan adaptation capabilities are realized through plan revision rules which are explained in section 3.4.2.

### 3.4.1 Properties of Revision Rules and Revisable Plans

The planning capabilities of SRCs are realized through policies that apply plan transformation rules to the SRP. These plan revision policies are triggered by local plan failures or exceptional situations that are observed and signaled during plan execution. The following three properties are important for plan revision.

***1. Invariance.*** *The properties of* SRC*s described above are invariant with respect to the application of plan transformation rules.* The empty SRC satisfies the conditions and, for all plan revision rules, "if $C$ then transform $P$ into $P'$" holds: if $P$ satisfies the required properties then $P'$ will satisfy them too.

*2. **Coverage.** A plan library of an* SRC *covers a set of states $\mathcal{S}$ if, for any state $s \in \mathcal{S}$, the plan library contains the plans ACHIEVE(s), MAINTAIN(s), BELIEF(s), and PERCEIVE(s).* The existence of these plans allows a very abstract formulation of the plan transformation rules. If the robot cannot achieve $s$ then the highest utility plan might be one that signals the corresponding failure.
*3. **Revisability.** A plan P is **revisable** if the* SRC *can repeatedly start executing P, interrupt the execution, and start it anew, with the resulting behavior of the robot then being very similar to the behavior displayed by executing the plan only once.* Restartability facilitates the smooth integration of plan revisions into ongoing activities: a partially executed restartable plan can be revised by terminating the plan and starting the new plan that contains the revisions (Beetz and McDermott, 1996).

The basics of how to specify such plan transformation rules and their application to SRPs are described and discussed in (Beetz and McDermott, 1997; Beetz and McDermott, 1994). Some of the rules work on predicted execution scenarios projected for the effects of SRPs (Beetz and Grosskreutz, 1998).

### 3.4.2 Revision Rules

The reasoning necessary to make an unanticipated decision and the installation of the decision is implemented as a plan revision rule that can be applied to the plan (Beetz and McDermott, 1997; Beetz and McDermott, 1994). Plan revision rules have the following form.

```
if    cond
then  transform  ips_1 at cp_1
      into       ops_1
      ...
      transform  ips_n at cp_n
      into       ops_n
```

where *cond* is the applicability condition, $ips_i$ at $cp_i$ the input plan schema and $ops_i$ the output plan schema of the rule. The applicability condition is a conjunction of literals. The input plan schema consists of a pattern variable to which the sub-plan with code path $cp_i$ is bound. $cp_i$ specifies the path in the syntax tree where the sub-plan to be revised can be found. The rule is applicable if the application condition holds. The resulting plan fragment $ops_i$ replaces $ips_i$. Thus, a transformation rule specifies a sequence of substitutions of new sub-plans for existing ones in the syntax tree.

Figure 3.13 shows the formalization of a revision rule that strategically reschedules office delivery tasks whenever the set of user commands changes. The rule revises the plan by adding another global policy that generates an unscheduled plan bug whenever an assumption underlying the current

**Plan Adaptation Rule** UNSCHEDULED-PLAN
**If** UNSCHEDULED-PLAN-BUG(?ACT-PRIMS)
 ∧ AT-LOC-PLANS(?ACT-PRIMS ?NAV-TASKS)
 ∧ BEST-SCHEDULE(?NAV-TASKS ?SCHED-CONSTRS)
 ∧ SCHEDULING-ASSMPTS (?NAV-TASKS ?SCHED-ASSMPTS))
**Then** **Transform**
 ?PRIM-ACTS **At** PRIMARY-ACTIVITIES
 **Into** **WITH-POLICY**
 **WHENEVER** ¬ ⋀(?SCHED-ASSMTS)
 SIGNAL(UNSCHEDULED-PLAN-BUG(?ACT-PRIMS))
 ?PRIM-ACTS
 **Transform**
 ?ACT-PRIMS **At** ACTIVE-PRIMARIES
 **Into** **PLAN** ?CMDS (?CONSTRS ∪ ?SCHED-CONSTRS)

**Fig. 3.13.** Revision rule UNSCHEDULED-PLAN. The condition part of the rule collects the *at location* sub-plans, computes a schedule for the *at location* sub-plans that aims at optimizing the route, and extracts the assumptions underlying the scheduling result. The **Then** part of the rule performs two transformations. The first one inserts a constraining sub-plan that signals a schedule bug whenever a schedule assumption is violated. The second one installs the ordering constraints that imply the schedule

schedule is detected as violated. The rule also revises the active primaries by adding the ordering constraints of the schedule to the constraints of the plan. The scheduling rule is applicable under a set of conditions specifying that (a) there is a bug of the category "unscheduled plan;" (b) the *at location* tasks contained in the active primaries are *?NAV-TASKS*; (c) *?SCHED-CONSTRS* are ordering constraints on *?NAV-TASKS* such that any order which satisfies *?SCHED-CONSTRS* will accomplish the active primary tasks quickly and avoid deadline violations and overloading problems; (d) *?NAV-TASKS* can be accomplished if *?SCHED-ASSTS* are satisfied.

Figure 3.14 shows the formalization of a adaptation rule that strategically reschedules office delivery tasks whenever the robot learns about an open door. The rule revises the plan by adding another global policy that generates an unscheduled plan bug whenever an assumption underlying the current schedule is detected as violated. The rule also revises the active primaries by adding the ordering constraints of the schedule to the constraints of the plan. The scheduling rule is applicable under a set of conditions specifying that (a) There is a belief change considering a door that was assumed to be closed but is indeed open; (b) The *at location* tasks contained in the active primaries are *?NAV-TASKS*; (c) *?SCHED-CONSTRS* are ordering constraints on *?NAV-TASKS* such that any order which satisfies *?SCHED-CONSTRS* will accomplish the active primary tasks fast and avoid deadline violations and overloading problems; (d) *?NAV-TASKS* can be accomplished if *?SCHED-ASSTS* are satisfied.

**Plan Adaptation Rule** UNSCHEDULED-PLAN-STRATEGY
**If** Belief-Change(Open-Door(?Room))
∧ Active-Primaries( ?Act-Prims)
∧ At-Loc-Plans( ?Act-Prims ?At-Loc-Tasks)
∧ Good-Schedule(?At-Loc-Tasks ?Sched-Constrs)
∧ Scheduling-Assmpts (?Nav-Tasks ?Sched-Assmpts))
**Then** **Transform** ?Prim-Acts **At** Primary-Activities
**Into** **WITH PLAN ADAPTOR**
**WHENEVER** ¬ ⋀( ?Sched-Assmts)
Signal(Unscheduled-Plan-Bug(?Act-Prims))
?Prim-Acts
**Transform** ?Act-Prims **At** Active-Primaries
**Into** **PLAN** ?Cmds (?Constrs ∪ ?Sched-Constrs)

**Fig. 3.14.** Strategic plan adaptation rule UNSCHEDULED-PLAN-STRATEGY. The difference to the tactical adaptation rule is the usage of the predicate GOOD-SCHEDULE that is realized through an anytime heuristic scheduling algorithm

The inference needed to decide whether there are instantiations for *?SCHED-CONSTRS* such that *GOOD-SCHEDULE* holds is more sophisticated. In fact, making this inference requires a planning process. The planning technique that is performed is called probabilistic prediction-based schedule debugging (PPSD) (Beetz, Bennewitz, and Grosskreutz, 1999). PPSD iteratively revises a given schedule until it has computed a satisfying schedule. The hill-climbing search that decides whether a new candidate schedule is better than a previous one is based on sampling a small number of possible execution scenarios for each schedule and therefore has a risk of being false. Even for sampling few execution scenarios, making one iteration requires several seconds. Thus, PPSD is too slow to be applied as a plan adaptation tactics.

Fig. 3.15 shows a tactical adaptation implemented as a revision rule. The rule is applied to "closed door" belief changes. It only runs for doors that have a high probability of being closed. The rule deletes the failed plan for the user command from the scheduled activities and adds it to the opportunities. Whenever the triggering condition of an opportunity is satisfied, it interrupts the active activity, completes its execution, and passes the control back to the interrupted activity. Note that all conditions of the adaptation tactics can be tested quickly.

An important property of the revision rules is that they keep the plans locational transparent and restartable. Both rules neither change *at location* sub-plans nor introduce new plans that are to be performed at particular locations. So the resulting plans are still locational transparent. Because the scheduling algorithm produces a schedule in which all completed plan steps are before the plan steps that remain to be carried out, the plans resulting from the rule Unscheduled-Plan-Strategy are restartable if the original plan was restartable. Finally, the Closed-Door-Tactics keeps the plan restartable under

```
Plan Adaptation Rule CLOSED-DOOR-TACTICS
If   Belief-Change(Open-Door(?Room))
     ∧ Usually-Closed(?Room) ∧ tlc-Name(?tlc ?tlc-Name)
     ∧ Delete-Act-Primary-tlc-Plan (?tlc-Name ?Remaining-tlc-Plans)
     ∧ Named-Subtask(?tlc-Name ?tlc-Task)
     ∧ rpl-Exp(?tlc-Task ?tlc-Plan)
Then Transform ?Act-Prims At Active-Primaries
     Into ?Rem-Act-Primaries
     Transform ?Opp-Prims At Opportunistic-Primaries
     Into TOP-LEVEL
            :TAG ?tlc-Name
                 SEQ WAIT-FOR (Open(?Room))
                     ?tlc-Plan
            ?Opp-Primaries
```

**Fig. 3.15.** Revision rule CLOSED-DOOR-TACTICS. If the robot fails to complete a delivery because of a closed door that is usually closed, then the heuristic adaptation tactics tells the robot to complete the delivery as soon as it learns that the door is open

the condition that the triggering condition becomes true while the plan gets executed.

In general, all plan adaptation rules that RHINO applies preserve locational transparency and restartability.

We have run our SRCs with a number of plan revision policies, including revisions for handling situations such as the case where the rooms where objects are to be picked up or delivered are closed, for installing schedules on navigation tasks in order to reduce execution time, for proposing museum tours for given time resources, for postponing opportunities if taking the opportunities might result in deadline violations, for integrating new commands, and for deleting completed commands.

The corresponding plan transformation rules are quite complex and could not work successfully without the SRCs satisfying the properties described before. The closed door revision, for example, extracts a sub-plan for a top-level command from the active primaries and installs the sub-plan as an opportunistic primary triggered by the observation that the corresponding door is open.

## 3.5 Related Work on Plan Representation

Let us now review the related work on plan representation. We will first survey different approaches for the integration of mechanisms into plan-based robot control. We will conclude with a comparison to other work on plan representation for autonomous robots.

**Related Work on Image Processing Plans.** Research in the active vision area (Aloimonos, Weiss, and Bandopadhay, 1987; Bajcsy, 1988; Ballard,

1991) focuses on increasing the performance and robustness of image processing routines by selectively applying special purpose routines that extract purposive information from the image. SRIPPs and RECIPE focus on particular aspects of active vision, namely how to specify sub-plans that can perform active image processing in a robot control/plan language and the resource management facilities necessary for robust performance under scarce resource constraints. Horswill (1995) studies specialization methods for deriving fast image processing methods from environmental constraints.

Ullman (1984)'s theory of visual routines proposes that our visual perception is achieved by modular visual routines. These are composed of simple visual actions using a small set of registers containing visual data. In conjunction with fluents, SRIPPs can analogously be considered to be an expressive high-level programming language for the implementation of vision routines.

RECIPE in conjunction with RPL and SRIPPs has much in common with the GARGOYLE system (Prokopowicz et al., 1996) embedded in a three-tiered robot control architecture (Bonasso et al., 1997). GARGOYLE is also a modular platform-independent image processing environment that can assemble and execute image processing pipelines. As with RECIPE, GARGOYLE has an interface to a high level robot control system but in the case of GARGOYLE, the image processing routines are not plans about which the control system can reason. RECIPE's two component design in which one robust, lightweight component acts as a system watchdog, restarting and reconfiguring failed processes emphasizes RECIPE's focus on overall system robustness. RECIPE is more specialized for the role of an image processing system running on a robot. Its modular architecture and run-time configurability give it a greater configurability than GARGOYLE and RECIPE is able to use this to good effect in conjunction with the RPL SRIPPs. There is a number of robot control systems that tightly integrate image processing modules e.g. (Alami et al., 1998) and (Crowley and Christensen, 1995). To the best of our knowledge, none of these systems are concerned with reasoning about image processing routines at a symbolic level.

**Related Work on Conversation Plans.** The problem of communicating with robots using natural language has been addressed in AI research from early on (cf. SHAKEY (Nilsson, 1984) and SHRDLU (Winograd, 1972)). More recent approaches can be found: the work of Torrance and Stein (1997) who studied the relation between natural language sentences and sensori-motor behavior for navigation tasks.

The idea of describing conversations in terms of speech acts that change the belief state of the listener is not new (Austin, 1962; Searle, 1969). AI planning techniques for generating plans for communicating information can also be found in the literature (for instance, (Cohen and Perrault, 1979)). Nevertheless, the most widely used techniques for specifying flexible communication are dialog grammars (Reichman, 1981).

To the best of our knowledge, the idea of using concurrent reactive control language for the specification of flexible dialog behavior is new. In this chapter we have extended the planning language RPL (McDermott, 1991) in such a way that it enables robots to lead restricted natural language dialogs and thereby carry out ambiguous commands and deal with situations that the robot cannot deal with by itself.

We have built our procedures for natural language interpretation and generation on top of implementations that were simple, well documented, and easily modifiable (Russell and Norvig, 1995; Norvig, 1992). As a result, neither the range of sentences that we can parse and interpret correctly nor the speed can compete with the "state of the art". Nevertheless, on average parsing takes less than three seconds which seems to be fast enough considering the time people require to read and answer their emails. Cole et al. (1997) give an excellent and extensive review of current state of natural language processing techniques.

Cole et al. (1997) classifies approaches to modeling dialog behavior into dialog grammars (e.g. (Reichman, 1981)), plan-based models of dialog (e.g. (Cohen and Perrault, 1979)), and joint action theories of dialog (e.g. (Lochbaum, Grosz, and Sidner, 1990)). Our work complements the research in this area in that it provides an appropriate implementation substrate for these approaches.

The TRAINS project is similar to our project in that it investigates the integration of communication capabilities into a problem-solving agent. However, while the TRAINS project focuses on the reasoning aspects of building a conversationally proficient problem-solving agent, we focus on the control and synchronization of conversational actions in a global problem solving context. The application of resource-adaptive computational methods to the generation and interpretation of conversational actions in dialogs (Weis, 1997) is investigated within the READY project. RHINO also assigns time resources to threads of control. However, it does not yet apply automated reasoning techniques for determining adequate distributions of resources.

**Related Work on Execution Time Planning.** Different solutions have been proposed for these kinds of decision problems. The first category of solutions tries to eliminate the need for making sophisticated execution time planning altogether. The idea is to precompute all possible decisions into a universal plan (Schoppers, 1987). This kind of approach has recently gained a lot of attention. A number of approaches transform planning problems into (partially observable) Markov decision problems ((PO)MDP) and compute the optimal actions for all possible states (Boutilier, Dean, and Hanks, 1998). The universal plan and (PO)MDP approaches, however, are only feasible if the state spaces are small or at least well structured.

The approaches that make execution time decisions that are not precomputed can be divided into those that only make situation-specific decisions without foresight and those that deliberate as much as necessary before they decide. Situation-specific decisions without foresight can be made very quickly (Firby, 1989). The problem, however, is that locally optimal decisions might be suboptimal in the global context. The second solution is to generate a complete plan for the new situation from scratch (Burgard et al., 1998). This "stop and replan" strategy is the best solution as long as planning is very fast. Often, however, determining plans with high expected utility under uncertainty will take seconds or even minutes whether or not it will result in a better course of action. The experiments in (Pollack and Ringuette, 1990) have shown that neither approach is optimal.

Thus, a number of more pragmatic solutions between these extremes have been proposed. One of these is to dynamically plan for only a small subset of the robot's goals. For example, Simmons et al. (1997b) only consider goals with the highest priority and tasks that can be accomplished on the way. This approach searches for local optima but fails to recognize global consequences of decisions. Layered architectures (Bonasso et al., 1997) run planning and execution at different levels of abstraction and with different time scales and employ a sequencing layer to synchronize between both levels. In these approaches the plans provide only guidelines for execution. Disadvantages of these approaches are that planning processes use models that are too abstract for predicting all consequences of the decisions they make. In addition, planning processes cannot exploit the control structures provided by the lower levels for specifying more flexible and reliable behavior. Yet other approaches make no commitment on how to make such decisions (Georgeff and Ingrand, 1989).

**Related Work on Plan Representation for Autonomous Robots.** Probably, the work that is most closely related to the work described in this chapter is the one by Firby and his colleagues. Firby (1992,1994) has developed concepts for the integration of physical and sensing actions into plan-based control language. Indeed, the RAP language developed by Firby (1987) is a precursor of the RPL plan language that is used in this thesis. Firby et al. (1996) also have described plans for a trash collecting application. There are a number of salient parallels between both approaches. These similarities include the use of concurrent sensor-driven control patterns, the integration of continuous control processes, monitoring and synchronization actions. A main difference between the work by Firby et al. (1996) and ours is that Firby has only used the plans for situation specific execution of sketchy plans, while in our case plan management operations project plans and revise them.

Other plan languages that provide many of the control patterns provided by our language include ESL and the plan language of PRS (Ingrand, Georgeff, and Rao, 1992). As in the case of RAP, PRS and ESL provide powerful interpreters but no sophisticated tools for predicting what will happen when a plan gets executed and for execution time plan revision.

There are plan representations for plan generation as well as plan execution. Wilkins and Myers (1995) propose a plan representation as an interlingua between a plan generation and plan execution system. Lyons and Hendriks (1992) has developed a planning approach for a plan language with similar expressiveness as RPL. The planning approach, however, has only been applied to highly repetitive tasks such as robot kitting.

## 3.6 Discussion

A plan notation, such as the one described in this chapter, is difficult to assess. In this section I will try to assess the significance of the language along the dimensions we have introduced in the beginning of the chapter: *representational adequacy*, *inferential adequacy*, *inferential efficiency*, and *acquisitional efficiency*. Unfortunately, these criteria are difficult to measure and even more so to compare the degree of satisfaction with the one achieved by other plan languages. We will explain in the following paragraphs in which respects these criteria are met by the proposed plan language.

The first thing to note is that we have proposed a application- and mechanism-specific plan language or, at least, to extend the plan language by application-specific plan schemata. The *image processing pipeline* and *at location* sub-plan macros are examples of specific plan language extensions. The use of these application-specific sub-plan macros greatly simplifies the tasks of plan management operations by representing frequent sub-plans modularly and transparently.

Besides the application-specific sub-plan macros we have introduced more general ones. Sub-plan macros such as *with stabilizers* and *with plan adaptors* are useful in applications in which plans that work under certain assumptions are much more efficient than plans that work in all situations. For example, in order to schedule office delivery requests it is more efficient to assume that the robot knows for every door whether it is open or closed and make a specific schedule for this situation. To perform the delivery tours reliably, the robot must be capable to reschedule its tour on the fly if its beliefs with respect to the status of a door changes. In general, to perform assumption-based plan execution and management the robot must be capable of detecting when assumptions are violated and to either restore the assumption or to revise the plan such that it does not need to make the assumptions any more.

The plans we are using are representational adequate in that they can specify behaviors that are flexible and reliable. They are also capable of expressing the control patterns necessary for the effective control of the robot's different mechanisms. In addition, they allow for the flexible interleaving of opportunities and recovery plans.

The plans are also inferentially adequate in that they allow for inferring information that is critical for successful plan management from the given plans. The inferential adequacy is achieved by extending concurrent reactive plans with declarative goals (Beetz and McDermott, 1992; Beetz, 2000) and using the XFRM-ML (Beetz and McDermott, 1997; Beetz,

2000), a concise and declarative notation for implementing general, commonsense plan revision methods as plan transformation rules. In addition, chapter 4 will describe and discuss a projection mechanism that predicts very realistically how the world will probably change as the plan gets executed.

Inferential efficiency is also mainly achieved through the properties of the plans. The transparency of the plans enables the plan management mechanisms to realize teleological inferences as pattern-directed retrievals. The generality of the plans implies that fewer decisions have to be made by the planning mechanisms and that prediction doesn't need to be very accurate. Finally, the interruptability facilitates the execution time revision of the plans.

Finally, the acquisitional efficiency is concerned with the difficulty to acquire the plans of robotic agents. Chapter 5 describes how a robotic agent can autonomously acquire symbolic navigation plans that can be used by the plan-based controller. It further explains how some of the notations we have introduced in this chapter help to make the learning process easier. Other advantages of the plan notation, such as the explicit and declarative representations for plan adaptors have not yet been exploited for the learning of plan libraries. We plan to investigate the automatic learning of plan adaptors in our future research.

# 4. Probabilistic Hybrid Action Models

In the last chapter we have seen how robotic agents can exhibit flexible and reliable behavior by using concurrent reactive plans. Performing competent plan management operations often requires autonomous robots to have foresight. Temporal projection, the computational process of predicting what will happen when a robot executes its plan, is essential for the robots to successfully plan their intended courses of action. Thus, the robot must be equipped with a temporal projection mechanism, an inference technique that predicts what will happen when the robot executes its plan.

To be able to project their plans, robots must have causal models that represent the effects of their actions. Most robot action planners use representations that include discrete action models and plans that define partial orders on actions. Therefore, they cannot automatically generate, reason about, and revise modern reactive plans. This has two important drawbacks. First, the planners cannot accurately predict and diagnose the behavior generated by their plans because they abstract away from important aspects of reactive plans. Second, the planners cannot exploit the control structures provided by reactive plan languages to make plans more flexible and reliable.

In this chapter we develop **phams** (Probabilistic Hybrid Action Models), action models that have the expressiveness for the accurate prediction of behavior generated by concurrent reactive plans. To the best of our knowledge, PHAMs are the only action representation used in action planning that provides programmers with means for describing the interference of simultaneous, concurrent effects, probabilistic state and action models, as well as exogenous events. PHAMs have been successfully used by an an autonomous robot office courier and a museum tour-guide robot to make predictions of full-size robot plans during the execution of these plans (Beetz, 2001).

This chapter makes several important contributions to the area of decision-theoretic robot action planning. First, we describe PHAMs, formal action models that allow for the prediction of the qualitative behavior generated by concurrent reactive plans. Second, we show how PHAMs can be implemented in a resource efficient way such that predictions based on PHAMs can be performed by robots while executing their plans. Third, we apply the plan projection method to probabilistic prediction-based schedule debugging and analyze it in the context of a robot office courier (Beetz, 2001).

The remainder of this chapter is organized as follows. Section 4.1 describes CRPs and the technical problems in the prediction of the physical robot behavior that they generate. Section 3 explains how reactive control processes and continuous change can be modeled using hybrid systems. Section 4 introduces McDermott's probabilistic rule language for the prediction of exogenous events and physical effects. We will also describe how the hybrid system model can be represented using the McDermott's rule language and show several important properties of our representation. Section 5 describes a much more efficient implementation of PHAMs that has been employed for prediction-based tour scheduling for an autonomous robot office courier. We conclude with an evaluation and a discussion of related work.

## 4.1 Projecting Delivery Tour Plans

At a certain level of abstraction, robot action planning systems consider plan steps as black boxes that generate behavior. The robot must have causal models of these plans that enable its planning system to predict what will happen when these steps get executed. We call these plan steps *low-level plans.*

In this section we consider a particular instance of low-level plans: low-level navigation plans. Navigation is the single most important action of autonomous mobile robots. Predicting the path that a robot will take is not only necessary to predict *where* the robot will be, even more importantly, it is also a prerequisite for predicting *what* the robot will be able to perceive. For example, whether the robot will perceive that a door is open depends on the robot taking a path that passes by the door, executing the door angle estimation routine while passing the door, and the door being within sensor range. Consequently, if the robot executes a plan step only if a door is open then in the end the execution of this plan step mainly depends on the actual path the robot will take. This implies that an action planning process must be capable of predicting the trajectory accurately enough in order to predict the global course of action correctly.

Navigation actions are representative for a large subset of the robots' physical actions: they are movements controlled by motors. Physical movements have a number of typical characteristics. First, they are often inaccurate and unreliable. Second, they cause continuous (and sometimes discontinuous) change of the respective part of the robot's state. Third, the interference of concurrent movements can often be described as the superposition of the individual movements.

To get better intuitions of the issues involved in realistically projecting navigation behavior, recall our description of the *low-level navigation plans* in chapter 3.2.1, which were used by an autonomous robot that has been successfully deployed as a museum tour-guide robot (Thrun et al., 2000) and performed office delivery tasks (Beetz, 2001). A low-level navigation plan specifies how the robot is to navigate from one location in its environment, its current position, to another one, its destination.

A low-level navigation plans consists of two components (see for a more detailed description). The first one specifies a sequence of target points to be sequentially visited by the robot. The second component specifies when and how the robot is to adapt its travel modes as it follows the navigation path. In many environments it is advantageous to adapt the travel mode to the surroundings: to drive carefully (and therefore slowly) within offices because offices are cluttered, to switch off the sonars when driving through doorways (to avoid sonar crosstalk), and to drive quickly in the hallways. Whenever the robot crosses the boundaries between regions it adapts the parameterization of the navigation system.

To discuss the issues raised by the projection of concurrent reactive plans, we sketch a delivery tour plan that specifies how an autonomous robot office courier is to deliver mail to the rooms A-113, A-111, and A-120 in figure 1 (Beetz, 2001). The mail for room A-120 has to be delivered by 10:30 (a strict deadline). Initially, the planner asked the robot to perform the deliveries in the order A-113, A-111, and A-120. Because the room A-113 is closed the corresponding delivery cannot be completed. Therefore, the planning system revises the overall plan such that the robot is to accomplish the delivery for A-113 as an opportunity. In other words, the robot will interrupt its current delivery to deliver the mail to A-113 (see figure 4.1) if the delivery can be completed.

```
with policy as long as in-hallway?
                whenever passing-a-door?
                    ESTIMATE-DOOR-ANGLE()
   with policy seq wait for open?(A-113)
                    DELIVER-MAIL-TO(DIETER)
        1. GO-TO(A-111)
        2. GO-TO(A-120) before 10:30
```

**Fig. 4.1.** Delivery tour plan with a concurrent monitoring process triggered by the continuous effects of a navigation plan (passing a door) and an opportunistic step. This concurrent reactive plans serves as an example for discussing the requirements that the causal models must satisfy

The plan contains constraining sub-plans such as "whenever the robot passes a door it estimates the opening angle of the door using its laser range finders" and opportunities such as "complete the delivery to room A-113 as soon as you learn the office is open". These sub-plans are triggered or completed by the continuous effects of the navigation plans. For example, the event passing a door occurs when the robot traverses a rectangular region in front of the door. We call these events ***endogenous events***.

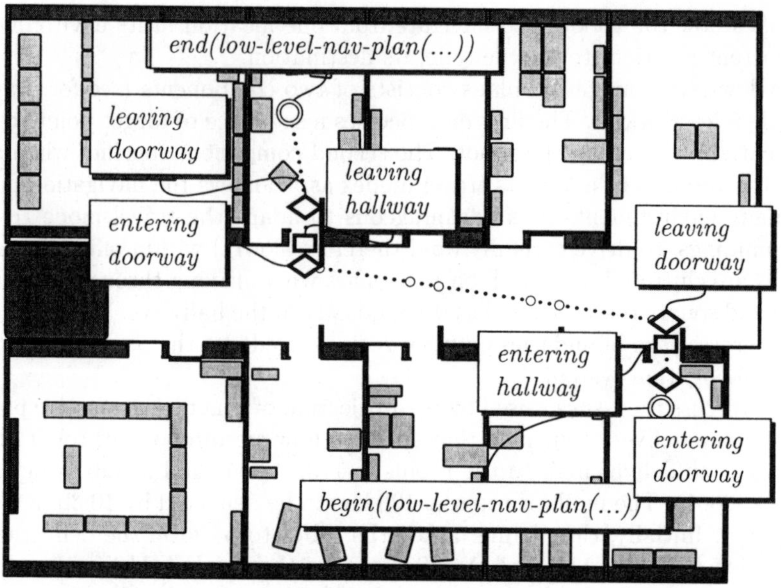


**Fig. 4.2.** Visualization of a projected execution scenario. The following types of events are depicted by specific symbols: change travel mode event by rhombus, start/stop passing doorway by small circles, start/stop low-level navigation plan by double circles, and entering doorway/hallway by boxes

Figure 4.2 shows a projected execution scenario for a low-level navigation plan embedded in the plan depicted in figure 3.9. The behavior generated by low-level navigation plans is modeled as a sequence of events that either cause qualitative behavior changes (e.g. adaptations of the travel mode) or trigger conditions that the plan is reacting to (e.g. entering the hallway or passing a door). The events depicted by rhomboids denote events where the CRP changes the direction and the target velocity of the robot. The squares denote the events entering and leaving offices. The small circles denote the events *starting* and *finishing passing a door*, which are predicted because a concurrent monitoring process estimates the opening angles of doors ***while*** the robot is passing them.

Such projected execution scenarios have been used for prediction-based debugging of delivery tours of an autonomous robot office courier. Beetz, Bennewitz, and Grosskreutz (1999) have shown that a controller employing predictive plan scheduling using the causal models described in this chapter can perform better than it possibly could without predictive capabilities (see also section 6.1).

There are several peculiarities in the projection of concurrent reactive plans that we want to point out here.

**Continuous Change.** Concurrent reactive plans activate and deactivate control processes and thereby cause continuous change of states such as the robot's position. The continuous change must be represented explicitly because CRPs employ sensing processes that continually measure relevant states (for example, the robot's position) and promptly react to conditions caused by the continuous effects (for example, entering an office).
**Reactive Control Processes.** Because of the reactive nature of robot plans, the events that have to be predicted for a continuous navigation process do not only depend on the process itself but also on the monitoring processes that are simultaneously active and wait for conditions that the continuous effects of the navigation process might cause. Suppose that the robot controller runs a monitoring process that stops the robot as soon as it passes an open door. In this case the planner must predict "robot passes door" events for each door the robot passes during a continuous navigation action. These events then trigger a sensing action that estimates the door angle, and if the predicted percept is an "open door detected" then the navigation process is deactivated. Other discrete events that might have to be predicted based on the continuous effects of navigation include entering and leaving a room, having come within one meter of the destination, etc.
**Interference between Continuous Effects.** Because the robots' effector control processes set voltages for the robot's motors, the possible modes of interference between control processes are limited. If they generate signals for the same motors the combined effects are determined by the so-called *task arbitration scheme* (Arkin, 1998). The most common task arbitration schemes are (1) behavior blending (where the motor signal is a weighted sum of the current input signals) (Konolige et al., 1997); (2) prioritized control signals (where the motor signal is the signal of the process with the highest priority) (Brooks, 1986); and (3) exclusion of concurrent control signals through the use of semaphores. In our plans, we exclude multiple control signals to the same motors but they can be easily incorporated in the prediction mechanism. Thus the only remaining type of interference is the superposition of movements such as turning the camera while moving
**Uncertainty.** There are various kinds of uncertainty and non-determinism in the robot's actions that a causal model should represent. It is often necessary to specify a probability distribution over the average speed and the displacements of points on the paths to enable models to predict the range of spatio-temporal behavior that a navigation plan can generate. Another important issue is to model probability distributions over the occurrence of exogenous events. In most dynamic environments exogenous events such as opening and closing doors might occur at any time.

## 4.2 Modeling Reactive Control Processes and Continuous Change

Let us now conceptualize the behavior generated by modern robot plans and the interaction between behavior and the interpretation of reactive plans. We base our conceptualization on the vocabulary of *hybrid systems*. Hybrid systems have been developed to design, implement, and verify embedded systems, collections of computer programs that interact with each other and an analog environment (Alur, Henzinger, and Wong-Toi, 1997; Alur, Henzinger, and Ho, 1996).

The advantage of a hybrid system-based conceptualization over state-based ones is that hybrid systems are designed to represent concurrent processes with interfering continuous effects. They also allow for discrete changes in process parameterization, which we need to model the activation, deactivation, and reparameterization of control processes through reactive plans. In addition, hybrid system-based conceptualizations can model the procedural meaning of *wait for* and *whenever* statements.

As pictured in figure 4.3, we consider the robot together with its operating environment as two interacting processes: the environment including the robot hardware, which is also called the controlled process, and the concurrent reactive plan, which is the controlling process. The state of the environment is represented by state variables including the variables $x$ and $y$, the robot's real position and $door\text{-}angle_i$ representing the opening angle of door $i$. The robot controller uses fluents to store the robot's measurements of these state variables (*robot-x*, *robot-y*, *door-a120*, etc.). The fluents are steadily updated by self-localization process and a model-based estimator for estimating the opening angles of doors. The control inputs of the plan for the environment process is a vector that includes the *travel mode*, the parameterization of the navigation processes and the current target point to be reached by the robot.

We will now model the controlled process as a hybrid system. Hybrid systems are continuous variable, continuous time systems with a phased operation. Within each phase, called *control mode*, the system evolves continuously according to the dynamical law of that mode, called *continuous flow*. Thus the state of the hybrid system can be thought of as a pair — the control mode and the continuous state. The control mode identifies a flow, and the continuous flow identifies a position in it. With each control mode there are also associated so-called *jump conditions*, that specify the conditions that the discrete state and the continuous state together must satisfy to enable a transition to another control mode. Once the transition occurs, the discrete state and the continuous state are changed abruptly. The *jump relation* specifies the valid settings of the system variables that might occur during a jump. Then, until the next transition, the continuous state evolves according to the flow identified by the new control mode.

When considering the interpretation of concurrent reactive plans as a hybrid system the control mode is determined by the set of active control

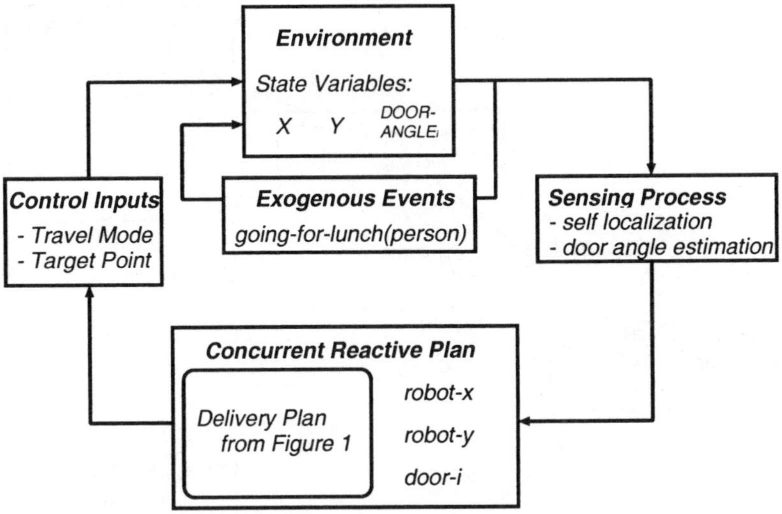


**Fig. 4.3.** The figure shows our conceptualization of the execution of navigation plans. The relevant state variables are the x and y coordinates of the robot's position and opening angles of the doors. The fluents that estimate these state variables are *robot-x*, *robot-y*, and *door-a120*

processes and their parameterization. The continuous state is characterized by the system variables $x$, $y$, and $o$ that represent the robot's position and orientation. The continuous flow describes how these state variables change as a result of the active control processes. This change is represented by the component velocities $\dot{x}$, $\dot{y}$, and $\dot{o}$. Thus for each control mode the robot's velocity is constant. Linear flow conditions are sufficient because the robot's paths can be approximated accurately enough using polylines (Beetz and Grosskreutz, 1998). They are also computationally much easier and faster to handle. The jump conditions are the conditions that are monitored by constraining control processes which activate and deactivate other control processes.

Thus the interpretation of a navigation plan according to the hybrid systems model works as follows. The hybrid system starts at some initial state $\langle cm_0, x_0 \rangle$. The state trajectory evolves with the control mode remaining constant and the continuous state $x$ evolving according to the flow condition of $cm$. When the continuous state satisfies the transition condition of an edge from mode $cm$ to a mode $cm'$ a jump must be made to mode $cm'$, where the mode might be chosen probabilistically. During the jump the continuous state may get initialized to a new value $x'$. The new state is the pair $\langle cm', x' \rangle$. The continuous state $x'$ evolves according to the flow condition of $cm'$.

Figure 4.4 depicts the interpretation of the first part of the navigation plan shown in figure 3.9. The interpretation is represented as a tree where

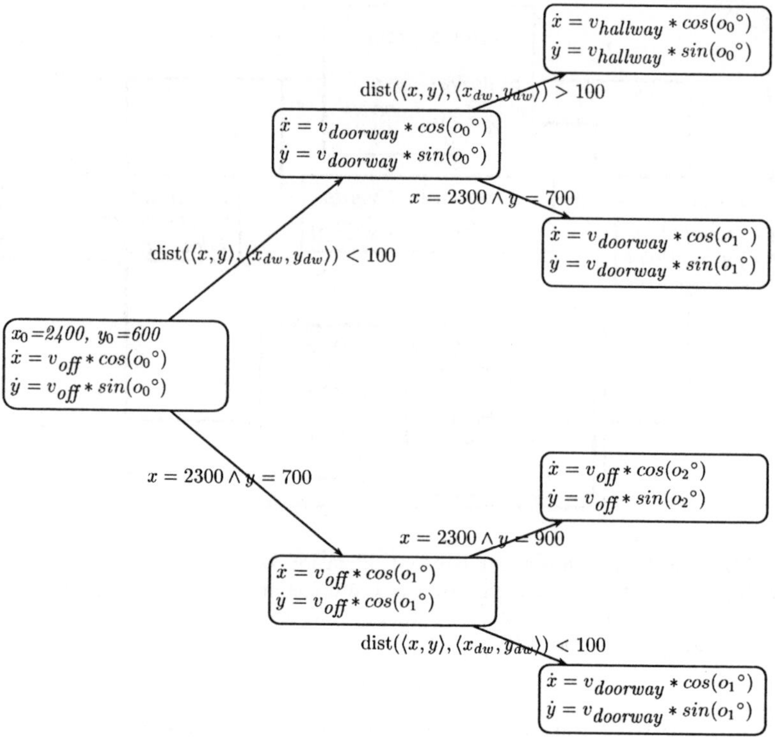


**Fig. 4.4.** The figure shows the hybrid automaton for the interpretation of the navigation plan pictured in figure 3.9. The possible control modes with the continuous flow equations are depicted as nodes and the mode transitions as arcs. The arcs are labeled with the jump conditions

the nodes represent the control modes of the corresponding hybrid system and the node labels the continuous flow. The edges are the control mode transitions labeled with the jump conditions. The robot starts executing the plan in room A-117. The initial control mode of the hybrid system is the root of the state tree depicted in figure 4.4. The initial state represents the state of computation where the first control processes of the two parallel branches are active, that is the processes for going to the intermediate target point **1** and maintaining the "office mode" as the robot's travel mode. The flow specifies that while the robot is in the initial control mode the absolute value of the derivative of the robot's position is $v_{off}$. The hybrid system transits into one of the subsequent states when either the first target point is reached or when the distance to the doorway becomes less than one meter. The transition condition for the upper edge is that the robot has come sufficiently close to the doorway, for the lower edge that it has reached the first target point. For the upper edge, the hybrid system goes into the state where the robot goes to the target point **2** while still keeping the office mode as its current travel mode. In the other transition the robot changes its travel mode to doorway and keeps approaching the first target point. The only variables that are changed through the control mode transitions are the velocity of the robot and its orientation. Both settings are implied by the flow condition of the respective successor states.

There is one issue that we have not yet addressed in our conceptualization: uncertainty. We model uncertainty with respect to the continuous effects and the achievement of jump conditions using multiple alternative successor modes with varying flows and jump conditions. We associate with the mode transitions a probability of their occurrence. This way we can, for example, represent rotational inaccuracies of navigation actions that are typical for mobile robots.

Let us now formalize our hybrid system conceptualization using a logical notation. To do so, we will use the following predicates to describe the evolution of their states: *jumpCondition(cm,e,c)*, *jumpSuccessor(e,cm',probRange)*, *jumpRelation(cm',* ***vals,flows****)*, and *probRange(e,max)*. *jumpCondition (cm,e,c)* represents that control mode *cm* is left along edge *e* when the condition *c* becomes true. A jump *e* causes the automaton to transit probabilistically into a successor mode.

For each possible successor we define a probability range *probRange*. For reasons that are explained below we represent the probability ranges such that they are non-overlapping, their relative sizes are proportional to the probability they represent, the sum of the ranges is 1, and that their boundaries have the form $\frac{i}{2^n}$, where $i$ and $n$ are integers. The predicate *probRange(e,*$2^n$*)* defines the sum of the ranges. A possible transition with a probability range $[\frac{i}{2^n}, \frac{j}{2^n}]$ is represented as *jumpSuccessor(e,cm',[i,j])*. The predicate *jumpRelation(cm',****vals****,* ***flows****)* means that upon entering control mode *cm'* the system variables and flows are initialized as specified by ***vals*** and ***flows***.

Using the predicates introduced above, we can state a probabilistic hybrid automaton for the interpretation of the plan depicted in figure 3.9 using the following facts.

*jumpRelation(*$cm_0$*,⟨2400,800⟩,⟨30,80⟩)*
*jumpCondition(*$cm_0$*,*$e_1$*,*$y \geq 900$*)*

*jumpSuccessor(*$e_1$*,*$cm_1$*,[1,12])*
*jumpSuccessor(*$e_1$*,*$cm_2$*,),[13,16])*
*probRange(*$e_1$*,16)*

*jumpRelation(*$cm_1$*,⟨ ⟩,⟨30,80⟩)*
*jumpRelation(*$cm_2$*,⟨ ⟩,⟨40,76⟩)*
...

The robot starts at position ⟨2400, 800⟩ in control mode $cm_0$ in which the robot leaves the lower office on the right. In this control mode the robot moves with 30cm/s into the x-direction and with 80cm/s in the y-direction. The navigation system leaves control mode $cm_0$ as soon as the y coordinate becomes greater than 900 by performing the transition $e_1$. If the system performs the transition $e_1$ then with 75% ($\frac{12}{16}$) probability the system transits into control mode $cm_1$ and with 25% ($\frac{4}{16}$) probability into the mode $cm_2$. The flow condition in $cm_1$ is $\dot{x} = 30$ and $\dot{y} = 80$ and in $cm_2$ $\dot{x} = 40$ and $\dot{y} = 76$.

To represent the state of a hybrid automaton we use the predicates *mode(cm)* and *startTime(cm,t)* to represent that the current control mode is *cm* and that *cm* started at time *t*. We use *flow(flow)* and *valuesAt*($t_i$,$\boldsymbol{val_i}$) to assert the flows and values of system variables for given time points. Further, the values of system variables can be inferred for arbitrary time points through interpolation on the basis of the current flow and the last instances of *valuesAt*($t_i$,$\boldsymbol{val_i}$). This is done using the predicate *stateVarsVals*:

$$\textit{stateVarVals}(\boldsymbol{vals}) \equiv \text{valuesAt}(t_0,\boldsymbol{vals_0}) \wedge \textit{now}(t) \wedge \textit{flow}(\boldsymbol{flow}) \wedge \boldsymbol{vals} = \boldsymbol{vals_0} + (t - t_0)\boldsymbol{flow}$$

where *now(t)* specifies that *t* is the current time. Note, in this conceptualization we represent the discrete state changes explicitly and the states within a mode using the mode's initial state and its flow. A particular state within a mode can be derived on demand using the predicate *stateVarVals.* Interference between different movements of the robot that are issued in different control threads are modeled through the mode's flow.

Figure 4.5 depicts an execution scenario, a possible evolution of the hybrid system, that represents how the execution of a robot controller might go. An execution scenario consists of a *timeline*, a linear sequence of events and their results. Timelines represent the effects of plan execution in terms of *time instants*, *occasions*, and *events. Time instants* are points in time at which the world changes due to an action of the robot or an exogenous event. Each time instant has a *date* which holds the time on the global clock at that particular time instant occurred. An *occasion* is a stretch of time over which a world state *P* holds and is specified by a proposition, which describes *P*, and the time interval for which proposition is true.

## 4.3 Probabilistic, Totally-Ordered Temporal Projection

In the last section we have seen how hybrid systems and execution scenarios are represented. In this section, we will see how we can predict execution scenarios from the specification of a hybrid system. For this purpose we use McDermott's rule language for probabilistic, totally-ordered temporal projection (McDermott, 1997). This rule language has the expressiveness needed for our purpose and a sound inference mechanism.

The different kinds of rules provided by this language are projection rules, effect rules, and exogenous event rules. We will describe these kinds of rules below.

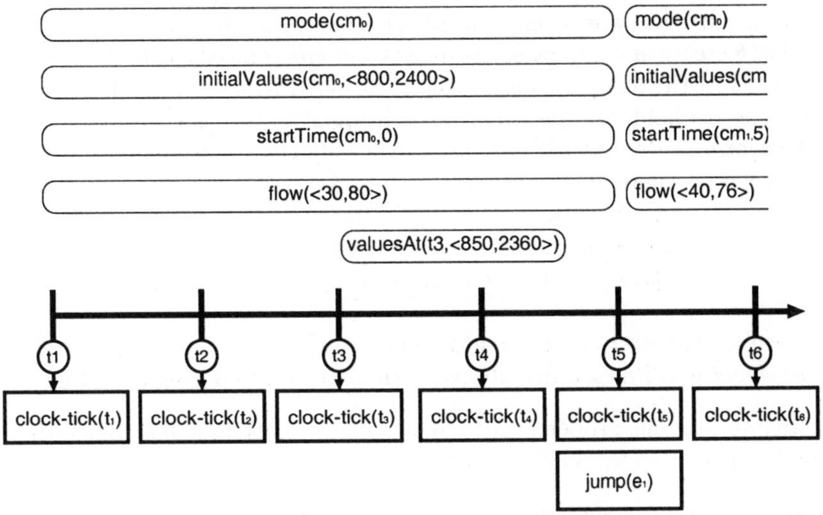


**Fig. 4.5.** Part of a timeline that represents a projected execution scenario for a low-level navigation plan. Time instants are depicted as circles, events as rectangles, and occasions as rectangles with round corners

- ***Projection rules*** can be used to specify the sequence of events caused by the interpretation of a low-level plan. Projection rules have the form

  <u>project rule</u> *name(args)*
  <u>if</u> *cond*
  <u>with a delay of</u> $\delta_1$ <u>occurs</u> $ev_1$
  ...
  <u>with a delay of</u> $\delta_n$ <u>occurs</u> $ev_n$

  and specify that if low-level plan *name(args)* is executed and condition *cond* holds then the low-level plan generates the events $ev_1$, ..., $ev_n$ with relative delays $\delta_1$, ..., $\delta_n$, respectively. Thus, projection rules generate a sequence of dated events.
  Uncertain models can be represented by sampling the $\delta_i$ from a probability distribution over durations and by specifying conditions that are satisfied only with a certain probability.
- ***Effect rules*** are used to specify conditional probabilistic effects of events. They have the form

  <u>e→p rule</u> *name*
  <u>if</u> *cond*
  <u>then</u> <u>with probability</u> $\theta$ <u>event</u> *ev* <u>causes</u> *effs*

and specify that whenever event *ev* occurs and *cond* holds, then with probability $\theta$ create and clip states as specified in *effs*. The effects of the e→p rule rules have the form $A$, causing the occasion $A$ to hold, clip $A$), causing $A$ to cease to hold, and persist $t$ $A$), causing $A$ to hold for $t$ time units.

- ***Exogenous event rules*** are used to specify the conditional occurrence of exogenous events. The rule

  p→e rule *name*
  while *cond* with an avg spacing of $\tau$ occurs *ev*

  specifies that over any interval in which *cond* is true, exogenous events *ev* are generated Poisson-distributed with an average spacing of $\tau$ time units.

Before proving properties of our model we must first introduce McDermott's semantics of possible worlds.

**Definition 4.1.** *A* ***world state*** *is a function from propositions to* $\{T,F,\perp\}$ *and is extended to boolean formulas in the usual way. An* **occurrence** *e@t is a pair* $c = (e,t)$*, where e is an event and t a time* ($t \in \Re_+$)*. An* ***occurrence sequence*** *is a finite sequence of occurrences, ordered by date. Its duration is the date of the last occurrence. A* ***world of duration L****, where* $L \in R_+$*, is a complete history of duration L, that is, it is a pair* $(C,H)$*, where C is an occurrence sequence with duration* $\leq L$*, and H is a function from* $[0,L]$ *to world states. In* $H(0)$ *all propositions are mapped to F, and if* $t1 < t2$ *and* $H(t1) \neq H(t2)$ *then there must be an occurrence e@t with* $t1 \leq t \leq t2$*.*

We use the following abbreviations: $(A \downarrow t)(W)$, "A after t in execution scenario W", to mean that there is some $\delta > 0$ such that $\forall t' : t < t' < t + \delta \Rightarrow H(t')(A) = T$. $(A \uparrow t)(W)$, "A before t in W" is similarly defined, but the upper bound for $t'$ includes t. $A \uparrow c$ and $A \downarrow c$ refer to the date of the occurrence $c$.

**Definition 4.2.** *If T is a set of rules as defined above, Exog is an occurrence sequence*[1]*, P is the set of propositions, and L is a real number* $\geq$ *duration(Exog), then an* ***L-Model of T and Exog*** *is a pair* $(U,M)$*, where U is a set of worlds of duration L such that* $\forall (C,H) \in U : Exog \subset C$*, and M is a probability measure on U that obeys the following restrictions:* $(A{\downarrow}t)$*,* $(A{\uparrow}t)$ *and e@t are considered as random variables.* $\overline{A}$ *is the "annihilation" of a conjunction A, that is the conjunction of the negations of the conjuncts of A.*

[1] *Exog* is an occurrence sequence, which represents the events generated by the interpretation of the robot's plan and modeled using projection rules.

1. ***Initial blank state:*** *$\forall A \in P : M(A \uparrow 0) = 0$.*
2. ***Event-effect rules:*** *If $T$ contains a rule instance*

   e→p rule *name* if *A* then with probability *r* event *e* causes *B*,

   *then for every date $t$, require that, for all nonempty conjunctions $C$ of literals from $B$: $M(C{\downarrow}t | E@t \wedge A{\uparrow}t \wedge \overline{B}{\uparrow}t) = r$.*
3. ***Event-effect rules when the events don't occur:*** *Suppose $B$ is an atomic formula, and let $R = \{R_i\}$ be the set of all instances of* e→p *rules whose consequents contain $B$ or $\neg B$. If* e→p rule *$R_i$ =* if *$A_i$* then with probability *$p_i$* event *$E_i$* causes *$C_i$, then let $D_i = A_i \wedge \overline{C_i}$, Then $M(B{\downarrow}t | B{\uparrow}t \wedge N) = 1$ and $M(B{\downarrow}t | \neg B{\uparrow}t \wedge N) = 0$ where $N = (\neg E_1@t \vee \neg D_1) \wedge (\neg E_2@t \vee \neg D_2) \wedge \ldots$.*
4. ***Event-occurrence rules:*** *For every time point $t$ such that no occurrence with date $t$ is in Exog and every event $E$, such that there is exactly one instance*

   p→e rule *name* while *A* with an avg spacing of *d* occurs *E*

   *with $M(a{\uparrow}t) > 0$ require*

$$lim_{dt\to 0} \frac{M(\text{occ. of class } E \text{ between } t \text{ and } t{+}dt | A{\uparrow}t)}{dt} = 1/d$$

$$lim_{dt\to 0} \frac{M(\text{occ. of class } E \text{ between } t \text{ and } t{+}dt | \neg A{\uparrow}t)}{dt} = 0$$

   *if there exists no so such rule, require*

$$lim_{dt\to 0} \frac{M(\text{occ. of class } E \text{ between } t \text{ and } t{+}dt)}{dt} = 0$$

5. ***Conditional independence:*** *If one of the previous clauses defines a conditional probability $M(\alpha|\beta)$, which mentions times $t$, then $\alpha$ is conditional independent, given $\beta$, of all other random variables mentioning times on or before $t$. That is, for arbitrary $\gamma$ mentioning times on or before $t$, $M(\alpha|\beta) = M(\alpha|\beta \wedge \gamma)$.*

McDermott (1997) shows that this definition yields a unique probability distribution $M$. He also gives a proof that his projection algorithm samples random execution scenarios sampled this unique probability distribution implied by the given probabilistic models.

### 4.3.1 Probabilistic Temporal Rules for PHAMs

In order to predict the evolution of a hybrid system, we specify rules in McDermott's rule language that, given the state of the system and a time *t*, predict the successor state at *t*. To predict the successor state, we must distinguish three cases: first, a control mode transition occurs; second, an exogenous event occurs; and third, the system changes according to the flow of the current mode.

We will start with the rules for predicting control mode jumps. To ensure that mode transitions are generated as specified by the probability distributions over the successor modes, we will use the predicate *randomlySampledSuccessorMode(e,cm)* and realize a random number generator using McDermott's rule language.

$$
\begin{aligned}
&randomlySampledSuccessorMode(e,cm) \equiv \\
&\quad probRange(e, max) \wedge randomNumber(n, max) \\
&\quad \wedge jumpSuccessor(e, cm, range) \wedge n \in range
\end{aligned}
$$

In order to probabilistically sample values from probability distributions we have to axiomatize a random number generator that asserts instances of the predicate *randomNumber(n,max)* that was used above (see (Beetz and Grosskreutz, 2000)). We do this by formalizing a *randomize* event. McDermott (1997) discusses the usefulness of, and the difficulties in realizing, nondeterministic exclusive outcomes. Therefore in his implementation he escapes to Lisp and uses a function that returns a random element.

**Lemma 4.1.** *At any time point randomNumber has exactly one extension randomNumber(r,max) where r is an unbiased random between* 0 *and max.*

<u>Proof:</u> Let *max** be the largest *probRange* extension and *randomBit(i,value)* the i-th random bit. The start event that causes the initial state timeline causes *randomBit(i,0)* $\forall 0 \leq i \leq \log max^*$. Thereafter, a *randomize* event is used to sample their value:

<u>e→p rule</u> RANDOMIZE
<u>if</u> *randomBit(i,val)* ∧ *negation(val,neg)*
<u>then</u> <u>with probability</u> *0.5*
    <u>event</u> *randomize*
    <u>causes</u> *randomBit(i,neg)* ∧ <u>clip</u> *randomBit(i,val)*

Rule MODE-JUMP causes a control mode transition as soon as the jump condition *cond* becomes true. The rule says that in any interval in which *cm* is the current control mode and in which the jump condition *cond* for leaving *cm* following edge *edge* a jump along *edge* will occur with an average delay of $\tau$ time units.

p→e rule MODE-JUMP
while *mode(cm)* ∧ *jumpCondition(cm,cond,edge)*
∧ *stateVarsVal(vals)* ∧ *satisfies(vals,cond)*
with an average spacing of $\tau$ *time units*
occurs *jump(edge)*

Rule JUMP-EFFECTS specifies the effects of an jump event on the control mode, system variables, and the flow. If *cm* is a control mode randomly sampled from the probability distribution over successor nodes for jumps along *edge* then the jump along *edge* has the following effects. The values of the state variables and the flow condition of the previous control mode $cm_{old}$ are retracted and the ones for the new control mode *cm* asserted.

e→p rule JUMP-EFFECTS
if *randomlySampledSuccessorMode(edge,cm)*
∧ *initialValues(cm,**val**)* ∧ *flowCond(cm,**flow**)* ∧ *now(t)*
∧ $mode(cm_{old})$ ∧ $flow(flow_{old})$ ∧ $valuesAt(t_{old},val_{old})$
then with probability *1.0*
event *jump(edge)*
causes *mode(cm)* ∧ *flow(**flow**)* ∧ *valuesAt(transTime,**val**)*
∧ clip $mode(cm_{old})$ ∧ clip $flow(flow_{old})$
∧ clip $valuesAt(t_{old},val_{old})$

Time is advanced using *clock-tick* events. With every CLOCK-TICK(*?t*) event the *now* predicate is updated by clipping the previous time and asserting the new one. Note, the time differs at most $dt_{clock}$time units from the actual time.

e→p rule CLOCK-RULE
if $now(t_o)$
then with probability *1.0*
event *clock-tick(t)*
causes *now(t)* ∧ clip $now(t_o)$

Exogenous events are modeled using rules of the following structure. When the navigation process is in the control mode *cm* and the values vals of the state variables satisfy the condition for the occurrence of the exogenous event *ev*, then The event *ev* occurs with average spacing of $\tau$ time units.

p→e rule CAUSE-EXO-EVENT
while *mode(cm)* ∧ *exoEventCond(cm,cond,ev)*
∧ *stateVarsVal(vals)* ∧ *satisfies(vals,cond)*
with an average spacing of $\tau$ *time units*
occurs *exoEvent(ev)*

The effects of exogenous event rules are specified by rules of the following form. The exogenous event *exoEvent(ev)* with effect specification *exoEffect(ev,**val**))* causes the values of the state variables to change from $\boldsymbol{val_o}$ to $\boldsymbol{val}$.

e→p rule EXO-EVENT-EFFECT
if *exoEffect(ev,**val**))* ∧ *valuesAt($t_o$,$val_o$)* ∧ *now(t)*
then with probability *1.0*
    event *exoEvent(ev)*
    causes *valuesAt(t,**val**)* ∧ clip *valuesAt($t_o$,$\boldsymbol{val_o}$)*

### 4.3.2 Properties of PHAMs

We have seen in the last section that a PHAM consists of the rules above and a set of facts that constitute the hybrid automata representation of a given CRP. In this section we investigate whether PHAMs make the "right" predictions.

There are essentially three properties of predicted execution scenarios that we want to ensure. First, predicted control mode sequences are consistent with the specified hybrid system. Second, mode jumps are predicted according to the specified probability distribution over successor modes. Third, between two successive events, the behavior is predicted according to the flow of the respective control mode.

Because McDermott's formalism does not allow for modeling instantaneous state transitions we can only show that control mode sequences in execution scenarios are probably approximately accurate. In our view, this is a low price for the expressiveness we gain through the availability of Poisson distributed exogenous events.

The subsequent lemma 4.2 states that control mode jumps can be predicted with arbitrary accuracy and arbitrarily high probability by decreasing the time between successive clock ticks.

**Lemma 4.2.** *For each probability $\epsilon$ and delay $\delta$, there exists a $\tau$ (average delay of the occurrence of an event after the triggering condition has become true) and a $dt_{clock}$ (time between two subsequent clock ticks) such that whenever a jump condition becomes satisfied, then with probability $\geq 1-\epsilon$ a jump event will occur within $\delta$ time units.*

***Proof:*** *Let $t$ be the time where the jump condition is fulfilled. If $\tau \leq \delta/(2\log(1/\epsilon))$ and $dt_{clock} \leq \delta/2$ then at most $\delta/2$ time units after $t$ the antecedent of rule* MODE-JUMP *is fulfilled. The probability that no event of class $jump(cm')$ occurs between $t+\delta/2$ and $t+\delta$ is $\leq e^{-\delta/(2\tau)} = e^{-log(1/\epsilon)} = \epsilon$, so with probability $\geq 1-\epsilon$ such an event will occur at most $\delta$ time units after $t$.*

□

This implies that there is always a non-zero chance that control mode sequences are predicted incorrectly. This happens when two jump conditions become true and the jump triggered by the later condition occurred before the other one. However, the probability of such incorrect predictions can be made arbitrarily small by the choice of $\tau$ and $dt_{clock}$.

The basic framework of hybrid systems does not take the possibility of exogenous events into account and thereby allows for proving strong system properties such as the reachability of goal states from arbitrary initial conditions or safety conditions for the system behavior (Alur, Henzinger, and Wong-Toi, 1997; Alur, Henzinger, and Ho, 1996). For the prediction of robot behavior in dynamic environments these assumptions, however, are unrealistic. Therefore, we only have a weaker property, namely that only between immediate subsequent events the predicted behavior corresponds to the flows specified by the hybrid system.

**Lemma 4.3.** *Let $W$ be an execution scenario, $e_1@t_1$ and $e_2@t_2$ be two immediate subsequent events of type jump or exoEvent, and cm be the control mode after $t_1$ in $W$. Then, for every occurrence $e@t$ with $t_1 < t \leq t_2$ $W(t)$(stateVarVals(**vals**)) is unique. Further, **vals** = **vals**$_1$ + $(t - t_1)$ * flow(cm), where **vals**$_1$ are the values of the state variables at $t_1$.*

***Proof:*** *There are only two classes of rules that affect the value of valuesAt and flow: rule* JUMP-EFFECTS, *and rule* EXO-EVENT-EFFECT. *These rules always clip and set exactly one extension of the predicates, thus together with the fact that the initial event asserts exactly one such predicate, the determined value is unique.*

*During the interval $t_1$ to $t_2$ the extension of stateVarVals evolves according to the flow condition of mode cm due to the fact that flow is not changed by rule* EXO-EVENT-EFFECT. *Thus it remains as initially set by rule* JUMP-EFFECTS, *which asserts exactly the flow corresponding to cm. The proposition then follows from the assumption of a correct axiomatization of addition and scalar-vector multiplication.*

□

Another important property of our representation is that jumps are predicted according to the probability distributions specified for the hybrid automaton.

**Lemma 4.4.** *Whenever a jump along an edge e occurs, the successor state is chosen according to the probability distribution implied by probRange and jumpSuccessor.*

***Proof:*** *This follows from the properties of the randomize event and Rule Jump-Effects.*

□

Using the lemmata we can state and show the central properties of PHAMs: (1) the predicted control mode transitions correspond to those specified by the hybrid automaton; and (2) the same holds for the continuous predicted behavior in between exogenous events; (3) Exogenous events are generated according to they probabilities over a continuous domain (this is shown in (McDermott, 1997)).

**Theorem 4.1.** *Every sequence of mode(cm) occasions follows a branch $(cm_i), ..., (cm_j)$ of the hybrid automaton.*

***Proof:*** *Each occasion mode(cm) must be asserted by rule* JUMP-EFFECTS. *Therefore there must have been a jump(e) event. Consequently, there must have been a jumpCondition from the previous control mode to cm.*

□

Because jump events are modeled as Poisson distributed events there is always the chance of predicting control mode sequences that are not valid with respect to the original hybrid system. So next we will bound the probability of predicting such mode sequences by choosing the parameterization of the jump event and clock tick event rules appropriately.

**Theorem 4.2.** *For every probability $\epsilon$ there exists an average delay of a mode jump event $\tau$ and a delay $dt_{clock}$ with which the satisfaction of jump conditions is realized such that with probability $\geq 1 - \epsilon$ the* **vals** *of stateVarVals occasions between two immediate subsequent exogenous event follow a state trajectory of the hybrid automaton.*

***Proof:*** *The proof is based on the property that jumps occur in their correct order with an arbitrary high probability. In particular, we can choose $\delta$ as a function of the minimal delay between jump conditions becoming true. Then, the jumps to successor modes occur with arbitrarily high probability (Lemma 2). Finally, according to Lemma 3 the trajectory of stateVarVals between transitions is accurate.*

□

## 4.4 The Implementation of PHAMs

We have now shown that PHAMs define probability distributions over possible execution scenarios with respect to a given belief state. The problem of using PHAMs is obvious. Nontrivial CRPs for controlling robots reliably require hundreds of lines of code. There are typically several control processes active, many more are dormant waiting for conditions that trigger their execution. The hybrid automata for such CRPs are huge, the branching factors for mode transitions are immense. Let alone the distribution of execution scenarios that they might generate. The accurate computation of this probability distribution is prohibitively expensive in terms of computational resources.

There is a second source of inefficiency in the realization of PHAMs. In PHAMs we have used clock tick rules, Poisson distributed events, that generate clock ticks with an average spacing of $\tau$ time units. We have done so, in order to formalize the operation of CRPs in a single concise framework. The problem with this approach is that in order to predict control mode jumps accurately we must choose $\tau$ to be very small. This, however, increases the number of clock tick events drastically and makes the approach infeasible for all but the most simple scenarios.

In order to draw sample execution scenarios from the distribution implied by the causal model and the initial state description we use an extension of the XFRM projector (McDermott, 1992b) that employs the RPL interpreter (McDermott, 1991) together with McDermott's algorithm for probabilistic temporal projection (McDermott, 1997). The projector takes as its input a CRP, rules for generating exogenous events, a set of probabilistic rules describing the effects of events and actions, and a (probabilistic) initial state description.

The plan projector works exactly like the plan interpreter, except whenever it runs across a low-level plan. In this case the interpreter interacts with the real world whereas the projector interacts with the timeline: it uses causal models of the low-level plans to predict the results of executing these plans and asserts their effects in the form of propositions on the timeline. Similarly, when the plan activates a sensor, the projector makes use of a model of the sensor and the state of the world as described by the timeline to predict the sensor reading.

In this section we investigate how we can make effective and informative predictions on the basis of PHAMs that can be performed at a speed that suffices for prediction-based online plan revision. To achieve effectiveness we use two means. First, we realize weaker inference mechanisms that are based on sampling execution scenarios from the distribution implied by the causal models and the initial state description. Second, we replace the clock tick event mechanism with a different mechanism that infers the occurrence of control mode jumps and uses the *persist* effect to generate the respective delay. We will detail these two mechanisms in the remainder of this section.

### 4.4.1 Projection with Adaptive Causal Models

Let us first turn to the issue of eliminating the inefficiencies caused by the clock tick mechanism. We will do so by replacing clock tick rules with a mechanism for tailoring causal models on the fly and using the *persist* effects of the probabilistic rule language.

For efficiency reasons the process of projecting a continuous process $p$ is divided into two phases. The first phase estimates a schedule for endogenous events caused by $p$ while considering possible effects of $p$ on other processes but not the effects of the other processes on $p$. This schedule is transformed into a context-specific causal model tailored for the plan which is to be projected. The second phase projects the plan $p$ using the model of endogenous events constructed in the first phase. This phase takes into account the interferences with concurrent events and revises the causal model if situations arise in which the assumptions of the precomputed schedule are violated.

The projection module uses a model of the dynamic system that specifies for each continuous control process the state variables it changes and for each state variable the fluents that measure that state variable. For example, consider the low-level navigation plans that steadily change the robot's position (that is the variables x and y). The estimated position of the robot is stored in the fluents *robot-x* and *robot-y*:

*changes(low-level-navigation-plan, x)*
*changes(low-level-navigation-plan, y)*
*measures(robot-x, x)*
*measures(robot-y, y)*

**Extracting Relevant Conditions.** When the projector starts projecting a low-level navigation plan it computes the set of pending conditions that depend on *robot-x* and *robot-y*, which are the fluents that measure the state variables of the dynamic system and are changed by the low-level navigation plan. These conditions are implemented as fluent networks.

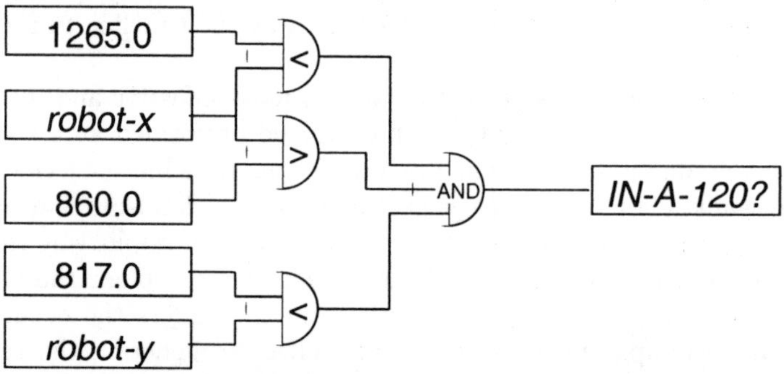


**Fig. 4.6.** Fluent network for being in room A-120. The robot believes to be in room A-120 if its estimated x-coordinate is between 860 and 1265 and the y-coordinate is smaller than 817, where the hallway begins

Fluent networks are digital circuits where the components of the circuit are fluents. Figure 4.6 shows a fluent network where the output fluent is true, if and only if the robot is in room A-120. The inputs of the circuit are the fluents *robot-x* and *robot-y* and the circuit is updated whenever *robot-x* and *robot-y* change.

Our reactive plans are set up such that the fluent networks that compute conditions for which the plan is waiting can be automatically determined using (PROLOG-like) relational queries:

<u>setof</u> *?fl-net ( fluent(?fl) ∧ status(?fl,pending)*
*∧ changes(low-level-nav-plan, ?state-var)*
*∧ measures(?state-var-fl, ?state-var)*
*∧ depends-on(?fl, ?state-var-fl)*
*∧ fluent-network(?fl, ?fl-net) )*
*?pending-fl-nets*

This query determines *?pending-fl-nets*, the set of fluent networks *?fl-net* such that *?fl-net* is a network with output fluent *?fl*. *?fl* causes a plan thread to pend and depends itself on a fluent measuring a state variable *?state-var* changed by the low-level navigation plan.

To predict when the fluent IN-A-120? will become true or false, we have to compute the region in the state space that corresponds to the fluent and compute the intersections of the robot's state trajectories with this region.

**Endogenous Event Schedules.** For each class of continuous processes we have to provide an *endogenous event scheduler* that takes the initial conditions of the process, the parameterization of the process, and the fluent networks that might be triggered by the process and computes the endogenous event schedule. The endogenous event scheduler for the low-level navigation plans is described in the next section. Given the kind of process (e.g., low-level navigation plan), the process parameters (e.g., the destination of the robot), and the pending fluent networks, the scheduler returns the sequence of predicted endogenous events, which are triples of the form $(\Delta t, \langle sv_1, ..., sv_n \rangle, \{ev_1, ..., ev_m\})$. $\Delta$t is the delay between the *i*th and the *i+1*th event in the schedule, $\langle sv_1, ..., sv_n \rangle$ the values of the state variables, and $\{ev_1, ..., ev_m\}$ the events that are to take place.

If a state for which the plan is waiting, becomes true at a time instance *t*, then at *t* a *passive-sensor-update* event is triggered. *passive-sensor-update* is an event model that takes a set of fluents as its parameters, retrieves the values of the state variables measured by these fluents, applies the sensor model to these values, and then sets the fluents accordingly.

**A Causal Model of Low-Level Navigation Plans.** Projecting the initiation of the execution of a navigation plan causes two events: the start event and a hypothetical completion event after infinite number of time units. This is shown in the following projection rule.

<u>*project rule*</u> LOW-LEVEL-NAVIGATION-PLAN
<u>*if*</u> *true*
<u>*with a delay of*</u> *0*
<u>*occurs*</u> *begin(low-level-nav-plan(?dest-descr, ?id, ?fluent)*
<u>*with a delay of*</u> $\infty$
<u>*occurs*</u> *end(low-level-nav-plan(?dest-descr, ?id, ?fluent)*

The effect rule of the start event of the low-level navigation plan computes the endogenous event schedule and asserts the schedule the occurrence of the next endogenous navigation event as an occasion to the timeline.

<u>*e→p rule*</u> ENDOGENOUS-EVENTS
<u>*if*</u> *endogenous-event-schedule(low-level-nav-plan(?dest-descr, ?schedule))*
<u>*then*</u> <u>*with probability*</u> *1.0*
    <u>*event*</u> *begin(low-level-nav-plan(?dest-descr, ?id, ?fluent))*
    <u>*causes*</u> *predicted-events(?id, ?schedule)*
        ∧ *running(robot-goto(?descr, ?id))*
        ∧ *next-nav-event(?id))*

The occasion *next-nav-event(?id)* triggers the next endogenous event *begin(follow-path(?here ⟨?x,?y⟩) ?dt ?id))*. The remaining two conditions determine the parameters of the *follow-path* event: the next scheduled event and the robot's position.

<u>*p→e rule*</u> CAUSE-EXO-EVENT
<u>*while*</u> *next-nav-event(?id)*
    ∧ *predicted-events(?id, ((?dt ⟨?x,?y⟩ ?evs) !?remaining-evs)*
    ∧ *robot-loc(?here)*
<u>*with an average spacing of*</u> *0.0001*
<u>*occurs*</u> *begin(follow-path(?here, ⟨?x,?y⟩, ?dt, ?id))*

The effect rule of the *begin(follow-path (...))* event specifies among other things that the next endogenous event will occur after *?dt* time units (<u>*persist*</u> *?dt sleeping(?id))*.

<u>*e→p rule*</u> FOLLOW-PATH
<u>*if*</u> *robot-loc(?coords)*
<u>*then*</u> <u>*with probability*</u> *1.0*
    <u>*event*</u> *begin(follow-path(?from, ?to, ?dt, ?id))*
    <u>*causes*</u> *running(follow-path(?from, ?to, ?dt, ?id))*
        ∧ <u>*clip*</u> *robot-loc(?coords)*
        ∧ <u>*clip*</u> *next-nav-event(?id)*
        ∧ <u>*persist*</u> *?dt sleeping(?id)*

If a running follow path event has finished sleeping the *end (follow-path (...))* event occurs.

<u>*p→e rule*</u> TERMINATE-FOLLOW-PATH
<u>*while*</u> <u>*not*</u> *sleeping(?id)*
    ∧ *running(follow-path(?from, ?to, ?time, ?id))*
<u>*with an average spacing of*</u> *0.0001*
<u>*occurs*</u> *end(follow-path(?from, ?to, ?time, ?id))*

Our model of low-level navigation plan presented so far suffices as long as nothing important happens while carrying out the plan. However, suppose that an exogenous event that causes an object to slip out of the robot's hand is projected at time instant *t* while the robot is in motion. To predict the new location of the object the projector predicts the location *l* of the robot at the time *t* and asserts it in the timeline.

Qualitative changes in the behavior of the robot caused by adaptations of the travel mode are described through e→p -rules. The following e→p -rule describes the effects of the event *nav-event(set-travel-mode(?n))*:

e→p rule SET-DOORWAY-MODE
if *travel-mode(?m)*
then with probability *1.0*
  event *nav-event(set-travel-mode(doorway))*
  causes clip *travel-mode(?m)*
    ∧ clip *obstacle-avoidance-with(sonar)*
    ∧ *travel-mode(doorway)*

The rule specifies that if at a time instant at which an event *nav-event(set-travel-mode(?n))* occurs the state *travel-mode(?m)* holds for some *?m*, then the states *travel-mode(?m)* and *obstacle-avoidance-with(sonar)* will (with a probability of *1.0*) not persist after the event has occurred, i.e., they are clipped by the event. The event causes the state *travel-mode(doorway)* to hold until it is adapted the next time.

### 4.4.2 Endogenous Event Scheduler

We have just shown how events are projected from a given endogenous event schedule, but we have not shown how the schedule is constructed. Thus, this section describes the endogenous event scheduler for low-level navigation plans. The scheduler predicts the effects of the low-level navigation plan on the state variables *x* and *y*. The endogenous event scheduler assumes the robot follows a straight path between the locations *1* to *5*. As we have pointed out earlier, there are two kinds of events that need to be predicted: the ones causing *qualitative physical change* and the ones causing the *trigger conditions* that the plan is waiting for.

The qualitative events caused by the low-level navigation plan pictured in figure 4.1 are the ones that occur when the robot arrives at the locations *1*, *2*, *3*, *4*, and *5* in which the robot either changes its travel mode or arrives at its destination. For each of these time instants the occurrence of a *set-travel-mode*-event is predicted.

The scheduler for trigger events works in two phases: (1) it transforms the fluent network into a condition that it is able to predict and (2) it applies an algorithm for computing when these events occur. The conditions that are caused by the low-level navigation plan can be represented as regions in the environment such that the condition is true if and only if the robot is within this region. The elementary conditions are numeric constraints on the robot's position or the distance of the robot to a given target point. The scheduler assumes that *robot-x* and *robot-y* are the only fluents in these networks that change their value during the execution of the plan. More complex networks can be constructed as conjunctions and disjunctions of the elementary conditions.

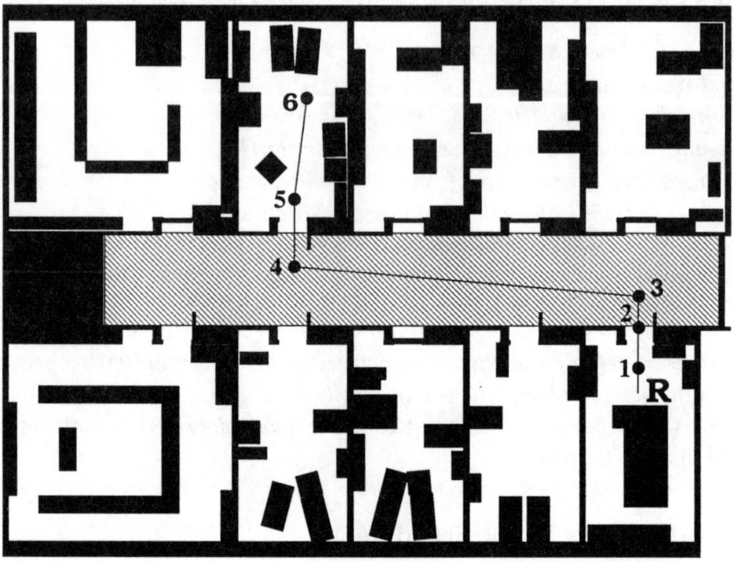


**Fig. 4.7.** Initially predicted endogenous events

In the next step the endogenous event scheduler overlays the straight line path through the intermediate goal points of the topological navigation path (see figure 4.1) with the regions computed in the previous step. It then computes a schedule for the endogenous events by following the navigation path and collecting the intersections with the regions (see figure 4.7). The result of the scheduling step is a sequence of triples of the form $(\Delta t_i, \langle x_i, y_i \rangle, \{ev_1, ..., ev_n\})$.

**Rescheduling Endogenous Events.** One problem that our temporal projector has to deal with is that a <u>*wait-for*</u> step might be executed while a low-level navigation plan is projected. For example, when the robot enters the hallway, the policy that looks for the opening angles of doors when passing them is triggered. Therefore, the causal model that was computed by the endogenous event scheduler is no longer sufficient. It fails to predict the "passing a door" events.

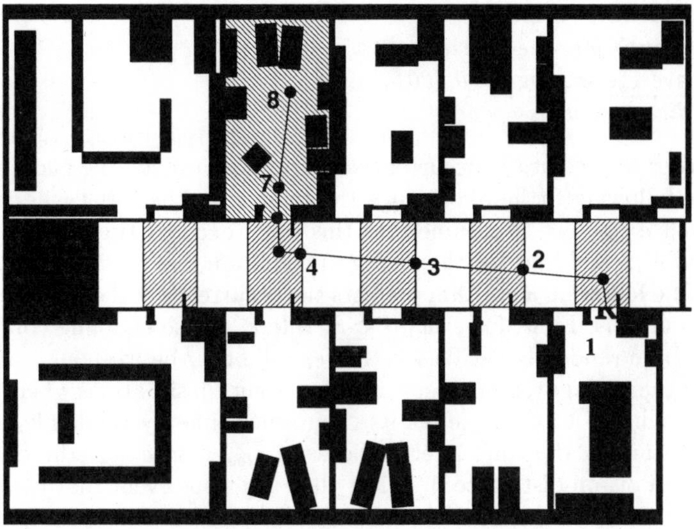


**Fig. 4.8.** Modified endogenous event schedule

These problems are handled by modifying the endogenous event schedule: whenever the robot starts waiting for a condition that is a function of the robot's position, it interrupts the projection of the low-level navigation plan, adapts the causal model of the low-level navigation plan, and continues with the projection. In the case of entering the hallway, a new endogenous event schedule is computed that contains endogenous events for passing doorways. This updated schedule of endogenous events is pictured in figure 4.8.

### 4.4.3 Projecting Exogenous Events, Passive Sensors, and Obstacle Avoidance

One type of exogenous event is an event for which we have additional information about their time of occurrence, such as the event that Dieter will be back from lunch around *12:25*. These kinds of events are represented by an <u>e→p</u> rule together with a <u>e→p</u> rule. The <u>e→p</u> rule specifies that the *start* event causes the state *before-dieters-door-opens()* to hold and persist for *?time* time units. The event *dieters-door-opens()* is triggered as soon as *before-the-door-opens()* no longer holds.

<u>e→p rule</u> BACK-FROM-LUNCH
<u>if</u> *about(?time, 12:25)* ∧ *difference(?time, *now*, ?wait-for))*
<u>then</u> <u>with probability</u> *1.0*
    <u>event</u> *start*
    <u>causes</u> <u>persist</u> *?wait-for before-the-door-opens*

p→e rule DOOR-OPENS
while thnot *before-the-door-opens*
with an average spacing of *0.0001*
occurs *dieters-door-is-opened*

In order to predict the occurrence of exogenous events, the plan projector does the following. It first computes the time when the robot will cause the next event $e_{next}$. Let us assume that this event occurs $t$ time units after the last event $e_{last}$ and that $c$ is the strongest condition that holds from $e_{last}$ until $e_{next}$.[2] The following algorithm predicts the occurrence of the next exogenous event accurately. First, for every P->E rule $r_i$ whose enabling condition is satisfied by $c$ randomly decide whether $e_i$ will occur between $e_{last}$ and $e_{next}$ based on the average temporal spacing of $e_i$ events in situations where $c$ holds. If $e_i$ is predicted to occur select its occurrence time by randomly selecting a time instant in the time interval between $e_{last}$ and $e_{next}$ (the exogenous events are Poisson distributed). Select the exogenous event that is predicted to occur the earliest, assert it to the timeline, and continue the projection after the occurrence of this event.

The last two components we need to describe are passive sensors, which are steadily updated and do not change the configuration of the robot, and the behavior of the collision avoidance routines.

Readings of passive sensors only need to be projected if the measured state variables change significantly or if the state variables traverse values that satisfy conditions for which the robot is waiting. For each of these situations there is an *update-passive-sensors* event.

Collision avoidance is not modeled except in situations in which the robot is told about objects that are moved around. In this case the endogenous event scheduler adds a region corresponding to the object. If the region blocks the way to the destination — that is the robot cannot move around the region — then a *possible-bump-event* is generated. The effect rule for a possible bump event specifies that, if the robot has activated sensors that can detect the object, the low-level navigation plan fails with a failure description "path blocked." Otherwise a bump event is generated. For example, since sonar sensors are the only sensors placed at table height, the collision avoidance module can avoid a collision with a table only if the sonar sensors are active. Thus, to predict a bump, the projector has to determine how long the sonar sensors have been switched off before the possible bump event occurs.

[2] The cases where enabling conditions of exogenous events are caused by the continuous effects between $e_{last}$ and $e_{next}$ are handled analogously to the achievement of triggering conditions.

### 4.4.4 Probabilistic Sampling-Based Projection

So far we have looked at the issue of efficiently predicting an individual execution scenario. We will now investigate the issue of drawing inferences that are useful for planning based on sampled execution scenarios.

Recently, probabilistic sampling-based inference methods have been proposed to infer information from complex distributions quickly and with bounded risk (Fox et al., 1999; Thrun, 1999). We will now discuss how we can use sampling-based projection for anticipating likely flaws with high probability.

Advantages of applying probabilistic sampling-based projection to the prediction of the effects of CRPs are that they work independently of the branching factor of the modes of the hybrid automaton and that it only constructs a small part of the complete PHAM.

But what kinds of prediction-based inferences can be drawn from samples of projected execution scenarios? The inference that we found most valuable for online revisions of robot plans is: do projected execution scenarios drawn from this distribution satisfy a given property $p$ with a probability greater than $\theta$? A robot action planner can use this type of inference to decide whether or not it should revise a plan to eliminate a particular kind of flaw: it should revise the plan if it believes that the flaws likelihood exceeds some threshold and ignore them otherwise. Of course, such inferences can be drawn based on samples only with a certain risk of being wrong. Suppose we want the planner to classify any flaw with probability greater than $\theta$ as to be eliminated and to ignore any flaw less likely than $\tau$. We assume that flaws with probability between $\tau$ and $\theta$ have no large impact on the robot's performance. How many execution scenarios should the plan revision module to classify flaws correctly with a probability greater than 95%?

A main factor that determines the performance of sample-based predictive flaw detection is the ***flaw detector***. A flaw detector classifies a flaw as to be eliminated if the probability of the flaw with respect to the robot's belief state is greater than a given threshold probability $\theta$. A flaw detector classifies a flaw as hallucinated if the probability of the flaw with respect to the robot's belief state is smaller than a given threshold $\tau$. Typically, we choose $\theta$ starting at 50% and $\tau$ smaller than 5%. We assume that the flaws with probability between $\tau$ and $\theta$ have no large impact on the performance of the robot. Because the classification is performed based on a set of samples, the classification has some risk of being incorrect.

Specific flaw detectors can be realized that differ with respect to (1) the time resources they require; (2) the reliability with which they detect flaws that should be eliminated; and (3) the probability that they hallucinate flaws. That is, they signal a flaw that is so unlikely that eliminating the flaw would decrease the expected utility.

To be more precise consider a flaw $f$ that occurs in the distribution of execution scenarios of a given scheduled plan with respect to the agent's

| | Prob. of Flaw $\boldsymbol{\theta}$ | | | | |
|---|---|---|---|---|---|
| | 50% | 60% | 70% | 80% | 90% |
| DET*(f,3,2)* | 50.0 | 64.8 | 78.4 | 89.6 | 97.2 |
| DET*(f,4,2)* | 68.8 | 81.2 | 91.6 | 97.3 | 99.6 |
| DET*(f,5,2)* | 81.2 | 91.3 | 96.9 | 99.3 | 99.9 |

**Fig. 4.9.** The table shows the probability of the flaw detectors DET*(f,i,2)* detecting flaws that have the probability $\boldsymbol{\theta}$ = 50%, 60%, 70%, 80%, and 90%

belief state with probability p. Further, let $X_i(f)$ represent the event that behavior flaw $f$ occurs in the $i$th execution scenario: $X_i(f) = 1$, if $f$ occurs in the $i$th projection and 0 otherwise.

The random variable $Y(f,n) = \sum_{i=1}^{n} X_i(f)$ represents the number of occurrences of the flaw $f$ in $n$ execution scenarios. Define a probable schedule flaw detector DET such that DET$(f,n,k)$ = *true iff* $Y(f,n) \geq k$, which means that the detector classifies a flaw $f$ as to be eliminated if and only if $f$ occurs in at least $k$ of $n$ randomly sampled execution scenarios.

Now that we have defined the schedule flaw detector, we can characterize it. Since the occurrence of schedule flaws in randomly sampled execution scenarios are independent from each other, the value of $Y(f)$ can be described by the binomial distribution $b(n,p)$. Using $b(n,p)$ we can compute the likelihood of overlooking a probable schedule flaw $f$ with probability $p$ in $n$ execution scenarios:

$$P(Y(f) < j) = \sum_{k=0}^{j-1} \binom{n}{k} * p^k * (1-p)^{n-k} \; .$$

Figure 4.9 shows the probability that the flaw detector DET$(f,n,2)$ for $n$ = 3,...,5 will detect a schedule flaw with probability $\boldsymbol{\theta}$. The probability that the detectors classify flaws less likely than $\tau$ as to be eliminated is smaller than 2.3% (for all $n \leq 5$).

When using the prediction-based scheduling as a component in the controller of the robot office courier we typically use DET*(f,3,2)*, DET*(f,4,2)*, and DET*(f,5,2)* which means a detected flaw is classified as probable if it occurs at least twice in three, four, or five detection readings.

Figure 4.10 shows the number of necessary projections to achieve $\beta = 95\%$ accuracy. For a detailed discussion see (Beetz, Bennewitz, and Grosskreutz, 1999).

The probabilistic sampling-based projection mechanism will become extremely useful for improving robot plans during their execution once the execution scenarios can be sampled fast enough. At the moment a projection takes a couple of seconds. The overhead is mainly caused by recording the interpretation of RPL plans in a manner that is far too detailed for our purposes. Through a simplification of the models we expect an immediate speed

| | θ | | | | | |
|---|---|---|---|---|---|---|
| | 1% | 10% | 20% | 40% | 60% | 80% |
| $\tau$ =.1% | 1331 | 100 | 44 | 17 | 8 | 3 |
| $\tau$ =1% | ⊥ | 121 | 49 | 17 | 8 | 3 |
| $\tau$ =5% | ⊥ | 392 | 78 | 22 | 9 | 3 |

**Fig. 4.10.** The table lists the number of randomly sampled projections needed to differentiate failures with an occurrence probability lower than $\tau$% from those that have a probability higher than $\theta$% with an accuracy of 95%

up of up to an order of magnitude. It seems that with a projection frequency of about 100 Hz one could start tackling a number of realistic problems that occur at execution time continually.

## 4.5 Evaluation

We have validated our causal model of low-level navigation plans and their role in office delivery plans with respect to computational resources and qualitative prediction results in a series of experiments. Section 6.1.5 shows the practical use of PHAMs for planning delivery tours of an autonomous robot office courier during their execution.

### 4.5.1 Generality

PHAMs are capable of predicting the behavior generated by flexible plans written in plan execution languages such as RAP (Firby, 1987) and PRS (Myers, 1996). To do so, we code the control structures provided by these languages as RPL macros. To the best of our knowledge PHAMs are the first realistic symbolic models of the sequencing layer of 3T architectures, the most commonly used software architectures for controlling intelligent autonomous robots (Bonasso et al., 1997). These architectures run planning and execution at different software layers and different time scales where a sequencing layer synchronizes between both layers. Each layer uses a different form of plan or behavior specification language. The planning layer typically uses a problem space plan, the execution layer employs feedback control routines that can be activated and deactivated. The intermediate layer typically uses a reactive plan language. The use of PHAMs would enable 3T planning systems to make more realistic predictions of the robot behavior that will be generated from their abstract plans. PHAMs are also capable of modeling different arbitration schemes and superpositions of the effects of concurrent control processes.

The causal models proposed here complement those described in (Beetz, 2000). Beetz (2000) describes sophisticated models of object recognition and manipulation that allow for the prediction of plan failures including ones that are caused by the robot overlooking or confusing objects, objects changing their location and appearance, and faulty operation of effectors. These models, however, were given for a simulated robot acting in a grid world. In this chapter, we have restricted ourselves to the prediction of behavior generated by modern autonomous robot controllers. Unfortunately, object recognition and manipulation skills of current autonomous service robots are not advanced enough for action planning. On the other hand, it is clear that action planning capabilities pay off much better if robots manipulate their environments and risk to do the wrong things.

### 4.5.2 Scaling Up

The causal models that we have described in section 4.4 have been used for execution time planning for a robot office courier. The plans that have been projected were the original plans for this application and typically several hundreds of code lines long. The projected execution scenarios contained hundreds of events. Because the projection of single execution scenarios can cost up to a second robots must revise plans based on very few samples. Thus, the robot can only detect probable flaws with high reliability.

Even with this severe limitation we were able to show that even with this preliminary implementation the robot can outperform controllers that lack predictive capabilities. The main source of inefficiency is the book keeping needed to reconstruct the entire computational state of the plan for any predicted time instant, an issue that we have not addressed in this chapter. Using a more parsimonious representation of the computational state we expect drastic performance gains.

### 4.5.3 Qualitatively Accurate Predictions

Projecting the plan listed in figure 4.1 generates a timeline that is about 300 events long. Many of these events are generated through rescheduling the endogenous events (21 times). figure 4.11 shows the predicted endogenous events (denoted by the numbered circles) and the behavior generated by the navigation plan in 50 runs using the robot simulator (we assume that the execution is interrupted in room A-111 because the robot realizes that the deadline can not be achieved). The qualitative predictions of behavior relevant for plan debugging are perfect. The projector predicts correctly that the robot will exploit the opportunity to go to location 5 while going from location 1 to 9.

The projector was also able to perfectly predict the relevant qualitative aspects of the other introductory example: whether or not the robot bumps into tables in front of door A-111.

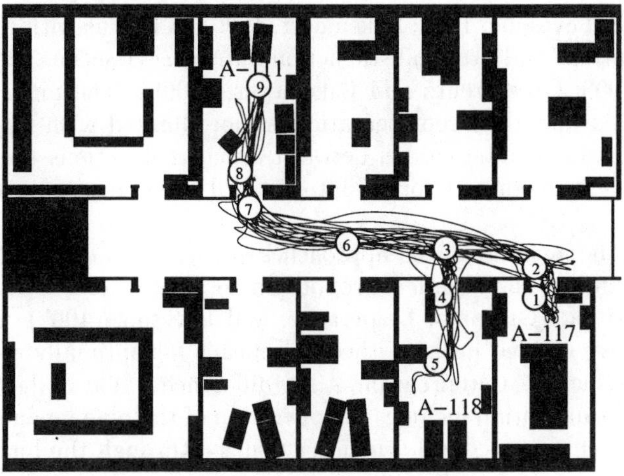


**Fig. 4.11.** The figure shows the trajectories of multiple executions of the navigation plan and the events that are predicted by the symbolic plan projector

## 4.6 Related Work on Temporal Projection

PHAMs represent external events, probabilistic action models, action models with rich temporal structure, concurrent interacting actions, and sensing actions in the domain of autonomous mobile robot control. There are many research efforts that formalize and analyze extended action representations and develop prediction and planning techniques for them. We know, however, only of approaches that address subsets of the aspects addressed by our representation. Related work comprises research on reasoning about action and change, probabilistic planning, numerical simulation, and qualitative reasoning.

*Reasoning about Action and Change.* Allen and Ferguson (1994) give an excellent and detailed discussion of important issues in the representation of temporally complex and concurrent actions and events. One important point that they make is that if actions have interfering effects then, in the worst case, causal models for all possible combinations of actions must be provided. In this paper, we have restricted ourselves to one kind of interference between actions: the transposition of movements which is the dominant kind of interference in physical robot behavior. In their article they do not address the issues of reasoning under uncertainty and efficiency with respect to computational resources.

A substantial amount of work has been done to extend the situation calculus (McCarthy, 1963) to deal with time and continuous change (Pinto, 1994; Grosskreutz and Lakemeyer, 2000a), exogenous (natural) actions (Reiter, 1996), complex robot actions (plans) (Levesque et al., 1997; Giacomo, Lesperance, and Levesque, 1997) using sensing to determine which action to execute next (Levesque, 1996; Lakemeyer, 1999) as well as with probabilistic state descriptions and probabilistic action outcomes (Bacchus, Halpern, and Levesque, 1999; Grosskreutz and Lakemeyer, 2000b). The main differences to our work is that their representation is more limited with respect to the kinds of events and interactions between concurrent actions they allow. In particular, we know of no effort to integrate all of these aspects at the same time.

Some of the most advanced approaches in this area are formalizations of various variants of the high-level robot control language GOLOG, in particular CONGOLOG (Giacomo, Lesperance, and Levesque, 1997). Boutilier et al. (2000) have applied decision theoretic means for optimally completing a partially specified GOLOG program. A key difference is that in the GOLOG approach the formalization includes the operation of the plan language whereas in our approach a procedural semantics realized through the high-level projector is used.

Hanks, Madigan, and Gavrin (1995) present a very interesting and expressive framework for representing probabilistic information, and exogenous and endogenous events for medical prediction problems. Because of their application domain they do not have to address issues of sophisticated percept-driven behavior as it is done in this chapter.

*Extensions to Classical Action Planning Systems.* Planning algorithms, such as SNLP (McAllester and Rosenblitt, 1991), have been extended in various ways to handle more expressive action models and different kinds of uncertainty (about the initial state and the occurrence and outcome of events) (Kushmerick, Hanks, and Weld, 1995; Draper, Hanks, and Weld, 1994; Hanks, 1990). These planning algorithms compute bounds for the probabilities of plan outcomes and are computationally very expensive. In addition, decision-theoretic action planning systems (see (Blythe, 1999) for a comprehensive overview) have been proposed in order to determine plans with the highest, or at least, sufficiently high expected utility (Haddawy and Rendell, 1990; Haddawy and Hanks, 1992; Williamson and Hanks, 1994). These approaches abstract away from the rich temporal structure of events by assuming discrete atomic actions and ignore various kinds of uncertainty.

Planning with action models that have rich temporal structure has also been investigated intensively (Allen et al., 1990; Dean, Firby, and Miller, 1988). IxTeT (Ghallab and Laruelle, 1994) is a planning system that has been applied to robot control and reasons about the temporal structure of plans to identify interferences between plan steps and resource conflicts. The planner/scheduler of the Remote Agent (Muscettola et al., 1998b) plans space maneuvers and experiments based on rich temporal causal models (Muscettola et al., 1998a; Pell et al., 1997b). A good overview of the integration of action planning and scheduling technology can be found in a recent overview article by Smith, Frank, and Jonsson (2000). So far they have considered uncertainty only with respect to the durations of actions.

Kabanza, Barbeau, and St-Denis (1997) model actions and behaviors as state transition systems and synthesize control rules for reactive robots from these descriptions. Their approach can be used to generate plans that satisfy complex time, safety, and liveness constraints. These approaches too are limited with respect to the temporal structure of the (primitive) actions being modeled and the kinds of interferences between concurrent actions that can be considered.

*MDP-Based Planning Approaches.* In recent years MDP (Markov decision process) planning has become a very active research field (Boutilier, Dean, and Hanks, 1998; Kaelbling, Cassandra, and Kurien, 1996). In the MDP approach robot behavior is modeled as a finite state automaton in which discrete actions cause stochastic state transitions. The robot is rewarded for reaching its goals quickly and reliably. A solution for such problems is a *policy*, a mapping from discretized robot states into, often fine-grained, actions.

MDPs form an attractive framework for action planning because they use a uniform mechanism for action selection and a parsimonious problem encoding. The action policies computed by MDPs aim at robustness and optimizing the average performance. A number of researchers have successfully considered navigation as an instance of Markov decision problems (MDPs) (Burgard et al., 2000; Kaelbling, Cassandra, and Kurien, 1996).

One of the main problems in the application of MDP planning techniques is to keep the problem encoding small enough so that the MDPs are still solvable. A number of techniques for complexity reduction can be found in the article written by Boutilier, Dean, and Hanks (1998). Yet, it is still very difficult to solve big planning problems in the MDP framework unless the state and action space is well structured.

Besides reducing the complexity of specifying models for, and solving MDP problems, extending the expressiveness of MDP formalisms is a very active research area. Semi Markov decision problems (Bradtke and Duff, 1995; Sutton, Precup, and Singh, 1999) add a notion of continuous time to the discrete model of change used in MDPs: transitions from one state to another one do no longer occur immediately, but according to a probability distribution. Others investigate mechanisms for hierarchically structuring MDPs (Parr and Russell, 1998), decomposing MDPs into loosely coupled subproblems (Parr, 1998), and making them programmable (Andre and Russell, 2001). Rohanimanesh and Mahadevan (2001) propose an approach for extending MDP-based planning to concurrent temporally extended actions. All these efforts are steps towards the kind of functionality provided in the PHAM framework they are still at an early stage of development. Another relationship between the research reported here and the MDP research is that the navigation routines that are modeled with PHAMs are implemented on top of MDP navigation planning. Belker, Beetz, and Cremers (2002) use the MDP framework to learning action models for the improved execution of navigation plans.

The application of MDP based planning to the reasoning through concurrent reactive plans is complicated by the fact that, in general, any activation and termination of a concurrent sub-plan might require a respective modification of the state and action space of the MDP.

Weaver (Blythe, 1995; Blythe, 1996) is another probabilistic plan debugger that capable of reasoning about exogenous events. Weaver uses Markov decision processes as its underlying model of planning. Weaver provides much of the expressiveness of PHAMs. Unlike Weaver, PHAMs are designed for reasoning about the physical behavior of autonomous mobile robots. Therefore, PHAMs add to Weaver's expressiveness in that they extensively support reasoning about concurrent reactive plans. For example, PHAMs can predict when the continuous effects of actions will trigger a concurrent monitoring process. PHAMs have built in the capabilities to infer the combined effects of two continuous motions of the robot.

*Qualitative Reasoning about Physical Processes.* Work in qualitative reasoning has researched issues in the quantization of continuous processes and focused among other things on quantizations that are relevant to the kind of reasoning performed. Hendrix (1973) points out the limitations of discrete event representations and introduces a very limited notion of continuous process as a representation of change. He does not consider the influence of multiple processes on state variables. Hayes (1985) represents events as *histories*, spatially bounded, but temporally extended, pieces in time space, and proposes that histories which do not intersect do not interact. In Forbus' Qualitative Process Theory (Forbus, 1984) a technique called limit analysis is applied to predict qualitative state transitions caused by continuous events. Also, work on simulation often addresses the adequacy of causal models for a given range of prediction queries, an issue that is neglected in most models used for AI planning. Planners that predict qualitative state transitions caused by continuous events include EXCALIBUR (Drabble, 1993).

## 4.7 Discussion

The successful application of AI planning to autonomous mobile robot control requires the planning systems to have more realistic models of the operation of modern robot control systems and the physical effects caused by their execution. In this chapter we have presented a ***probabilistic hybrid action models (*phams*)***, which are capable of representing the temporal structure of continuous feedback control processes, their non-deterministic effects, several modes of their interferences, and exogenous events. We have shown that PHAMs allow for predictions that are, with high probability, qualitatively correct. We have also shown that powerful prediction-based inferences such as deciding whether a plan is likely to cause a flaw with a probability exceeding a given threshold can be drawn fast and with bounded risk. Finally, we have demonstrated using experiments carried out on a real robot and a robot simulator that robots that decide based on predictions generated from PHAMs can avoid execution failures that robots without foresight cannot.

We believe that equipping autonomous robot controllers with concurrent reactive plans and prediction-based online plan revision based on PHAMs is a promising way to improve the performance of autonomous service robots through AI planning both significantly and substantially.

# 5. Learning Structured Reactive Navigation Plans

So far, we have described how the plans of robotic agents can be represented in order to make competent plan management operations feasible under strong resource constraints and how the robot can predict what might happen when it executes its plans. We have not yet addressed the issue of how the plan schemata could be acquired that are needed by the robot.

This chapter describes XFRMLEARN, a system that learns structured symbolic navigation plans automatically. Given a navigation task, XFRMLEARN learns to structure continuous navigation behavior and represents the learned structure as compact and transparent plans. The structured plans are obtained by starting with monolithic default plans that are optimized for average performance and adding sub-plans to improve the navigation performance for the given task. Compactness is achieved by incorporating only sub-plans that achieve significant performance gains. The resulting plans support action planning and opportunistic task execution.

In the remainder of the chapter we proceed as follows. The next section characterizes navigation path planning as a form of Markov decision problem. Section 5.2 gives an overview of the XFRMLEARN learning system. After this section we recapitulate the representation of structured reactive navigation plans. The subsequent section details the operation of XFRMLEARN. We conclude with our experimental results and a discussion of related work.

## 5.1 Navigation Planning as a Markov Decision Problem

A number of researchers consider navigation as an instance of Markov decision problems (MDPs) (Thrun et al., 1998; Kaelbling, Cassandra, and Kurien, 1996). They model the navigation behavior as a finite state automaton in which atomic navigation actions cause stochastic state transitions. The robot is rewarded for reaching its destination quickly and reliably. A solution for such problems is a *policy*, a mapping from discretized robot poses into fine-grained navigation actions.

MDPs form an attractive framework for navigation because they use a uniform mechanism for action selection and a parsimonious problem encoding. The navigation policies computed by MDPs aim at robustness and optimizing the average performance. One of the main problems in the application of

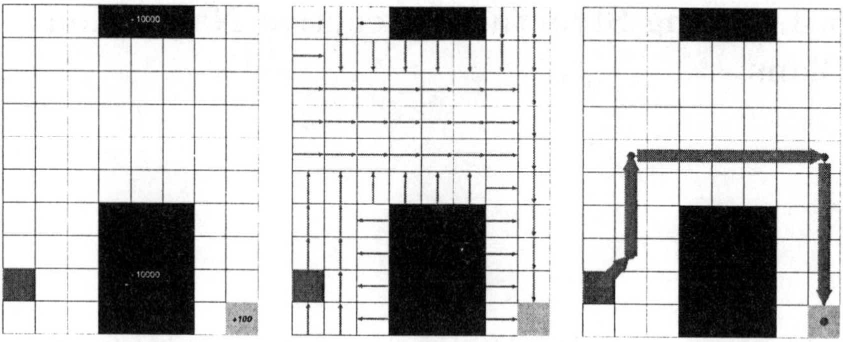


(a) (b) (c)

**Fig. 5.1.** Navigation planning as a Markov decision problem. Sub-figure (a) shows the navigation problem in a simple grid world. The start position is depicted by the dark grey and the destination by the light grey cell. Cells occupied by an obstacle are depicted as black. The agent receives a reward when reaching the destination, is penalized for hitting an obstacle, and has to pay some cost for executing an action. Sub-figure (b) shows for each grid cell the optimal action to execute. Sub-figure (c) the optimal path from a given position

MDP planning techniques is to keep the state space small enough so that the MDPs are still solvable. This limits the number of contingent states that can be considered.

In this chapter we investigate the problem of learning structured reactive navigation plans (SRNPs) from executing MDP navigation policies. The specification of good task and environment-specific SRNPs, however, requires tailoring their structure and parameterizations to the specifics of the environmental structures and empirically testing them on the robot. Our thesis is that a robot can autonomously learn compact and well-structured SRNPs by using MDP navigation policies as default plans and repeatedly inserting sub-plans into the SRNPs that significantly improve the navigation performance. This idea works because the policies computed by the MDP path planner are already fairly general and optimized for average performance. If the behavior produced by the default plans were uniformly good, making navigation plans more sophisticated would be of no use. The rationale behind requiring sub-plans to achieve significant improvements is to keep the structure of the plan simple.

## 5.2 An Overview on XFRMLEARN

The integration of XFRMLEARN into the plan-based controller is best visualized as shown in figure 5.2. XFRMLEARN is applied to the RHINO navigation system (Thrun et al., 1998), which has shown impressive results in several longterm experiments (Thrun et al., 1999). Conceptually, this robot navigation system works as follows. A navigation problem is transformed into a Markov decision problem to be solved by a path planner using a value iteration algorithm. The solution is a policy that maps every possible location into the optimal heading to reach the target. This policy is then given to a reactive collision avoidance module that executes the policy taking the actual sensor readings and the dynamics of the robot into account (Fox, Burgard, and Thrun, 1997).

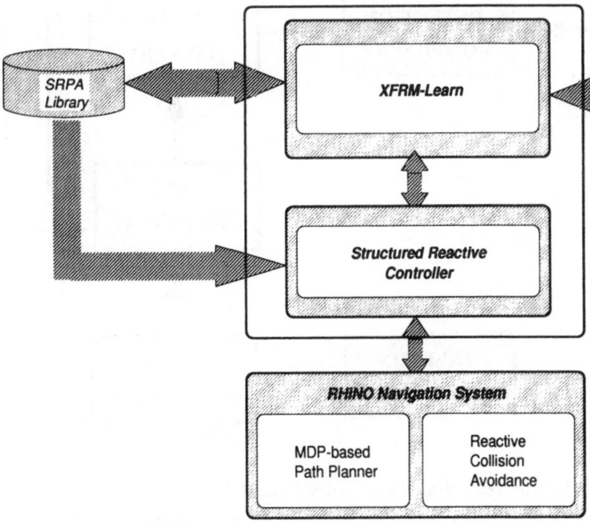


**Fig. 5.2.** Integration of XFRMLEARN into the plan-based controller. XFRMLEARN uses a knowledge base that contains knowledge for diagnosing conspicuous navigation behavior and plan revision methods for producing plans that exhibit better performance. XFRMLEARN incrementally improves the navigation plans in the plan library. The plans are used by the structured reactive controller to perform its application tasks. The SRC controls the RHINO navigation system that consists of an MDP path planner and a collision avoidance module

The RHINO navigation system can be parameterized in different ways. The parameter PATH is a sequence of intermediate points which are to be visited in the specified order. COLLI-MODE determines how cautiously the robot should drive and how abruptly it is allowed to change direction. The parameter SONAR specifies whether the sonar sensors are to be used in addition to laser sensors for obstacle detection.

The navigation system reactively selects a rotational and translational velocity to approach the next point on the robot's intended path, which is optimal with respect to its utility function. COLLI-MODE specifies a term that weighs the importance of the (1) clearance of the intended path, (2) translational velocity and (3) pose at the end of the path (Fox, Burgard, and Thrun, 1997).

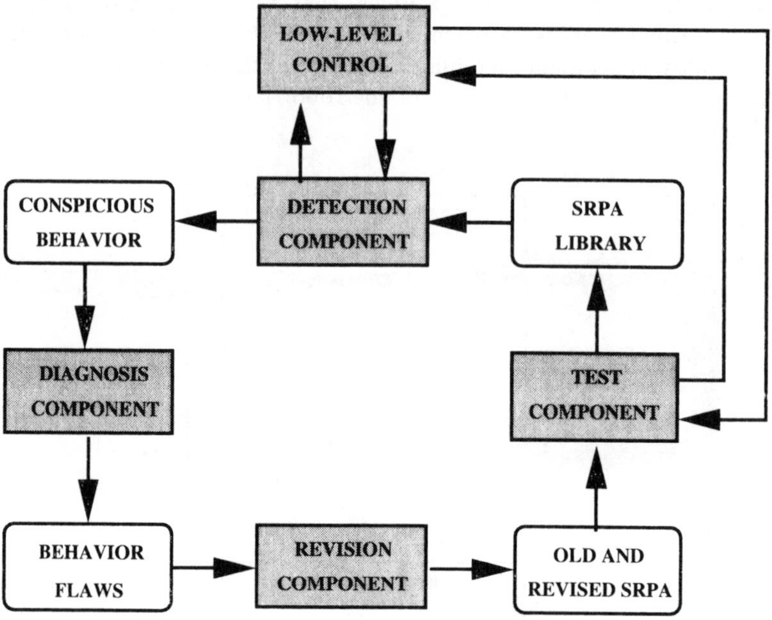


**Fig. 5.3.** XFRMLEARN's "detect, diagnose, revise, and test" cycle. The detection step selects a plan from the library, executes it and diagnoses conspicuous behaviors stretches. The diagnosis step tries to explain the suspicious stretches using its diagnosis knowledge. The diagnoses are then used to index revision rules to produce plans that exhibit better behavior. The best revision is applied. The revised and the original plan are then tested against each other to decide whether the revised plan should be substituted in the plan library for the original one

RHINO's navigation behavior can be improved because RHINO's path planner solves an idealized problem that does not take the desired velocity, the dynamics of the robot, the sensor crosstalk, and the expected clutteredness fully into account. The reactive collision avoidance component takes these aspects into account but makes only local decisions.

XFRMLEARN executes a "detect, diagnose, revise, and test" for learning SRNPs (see Figure 5.3). XFRMLEARN starts with a default plan that transforms a navigation problem into an MDP problem and passes the MDP problem to RHINO's navigation system. After RHINO's path planner has determined the navigation policy the navigation system activates the collision avoidance module for the execution of the resulting policy. XFRMLEARN records the resulting navigation behavior and looks for stretches of behavior that could be possibly improved. XFRMLEARN then tries to explain the improvable behavior stretches using causal knowledge and its knowledge about the environment. These explanations are then used to index promising plan revision methods that introduce and modify sub-plans. The revisions are then tested in a series of experiments to decide whether they are likely to improve the navigation behavior. Successful sub-plans are incorporated into the symbolic plan.

## 5.3 Structured Reactive Navigation Plans

Let us briefly recapitulate our representation of navigation plans 3.3.2. SRNPs have the following syntax.

<u>*navigation plan*</u> *(s,d)*
    <u>*with subplans*</u> *subplan**
  DEFAULT-GO-TO ( *d* ).

with *subplans* of the form

SUBPLAN-CALL(*args*)
    <u>*parameterizations*</u> $\{\ p \leftarrow v\ \}^*$
    <u>*path constraints*</u> $\{\ \langle\ x,y \rangle\ \}^*$
    <u>*justification*</u> *just*

where $p \leftarrow v$ specifies a parameterization of the subsymbolic navigation system. In this expression $p$ is a parameter of the subsymbolic navigation system and $v$ is the value $p$ is to be set to. The parameterization is set when the robot starts executing the subplan and reset after the subplan's completion. The path constraints are sequences of points that specify constraints on the path the robot should take when carrying out the subplan. The subplan call together with its arguments specifies when the subplan starts and when it terminates (see below). The last component of the subplan is its justification, a description of the flaw the subplan is intended to eliminate. To be more concrete consider the following SRNP, which is depicted in figure 5.10b).

<u>*navigation plan*</u> *(desk-1,desk-2)*
    <u>*with subplans*</u>
      TRAVERSE-NARROW-PASSAGE($\langle 635, 1274\rangle$,$\langle 635, 1076\rangle$)
        <u>*parameterizations*</u> *sonar* ← *off*
            *colli-mode* ← *slow*
        <u>*path constraints*</u> $\langle 635, 1274\rangle$,$\langle 635, 1076\rangle$
        <u>*justification*</u> *narrow-passage-bug-3*
      TRAVERSE-NARROW-PASSAGE(...)
      TRAVERSE-FREE-SPACE(...)
  DEFAULT-GO-TO ( *desk-2* )

The SRNP above contains three subplans: one for leaving the left office, one for entering the right one, and one for speeding up the traversal of the hallway. The subplan for leaving the left office is shown in more detail. The path constraints are added to the plan for causing the robot to traverse the narrow passage orthogonally with maximal clearance. The parameterizations of the navigation system specify that the robot is asked to drive slowly in the narrow passage and to only use laser sensors for obstacle avoidance to avoid the hallucination of obstacles due to sonar crosstalk.[1]

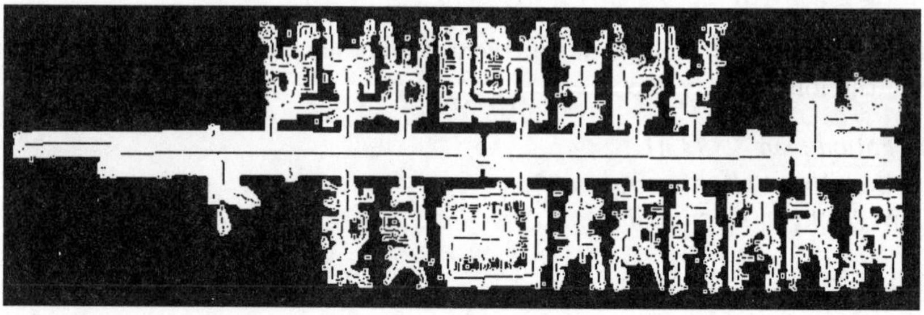

**Fig. 5.4.** A learned environment map with its main axes Voronoi skeleton. The map is a finegrained occupancy gridmap. The black pixels denote grid cells that are occupied by an obstacle. The picture also contains the main axes Voronoi diagram (see text)

The plan interpreter expands the symbolic navigation plan macros into procedural reactive control routines. Roughly, the plan macro expansion does the following. First, it collects all path constraints from the subplans and passes them as intermediate goal points for the MDP navigation problem. Second, the macro expander constructs for the subplan call a monitoring process that signals the activation and termination of the subplan. Third, the following concurrent process is added to the navigation routine: wait for the activation of the subplan, set the parameterization as specified, wait for the termination of the subplan, and reset the parameterization.

## 5.4 XFRMLEARN in Detail

In this Section we will describe the three learning steps, ***analyze***, ***revise***, and ***test*** in more detail.

[1] The programmers can specify in which areas it is safe to ignore sonar readings for obstacle avoidance.

### 5.4.1 The "Analyze" Step

A key problem in learning structured navigation plans is to structure the navigation behavior well. Because the robot must start the subplans and synchronize them, it must be capable of "perceiving" the situations in which subplans are to be activated and deactivated. Besides being perceivable, the situations should be relevant for adapting navigation behavior. We use the concept of *narrowness* for detecting the situations in which the navigation behavior is to be adapted. Based on the concept of narrowness the robot can differentiate situations such as free space, traversing narrow passages, entering narrow passages, and leaving narrow passages.

**Fig. 5.5.** A CAD map with its main axes Voronoi skeleton. The distance of the main axes Voronoi skeleton to the next obstacle is taken as a measure of how much clearance the robot could have in principle when traversing some given region

We compute the perceived *narrowness* as a function of the robot's position. To do so, we use a *main axes Voronoi skeleton*, a particular environment representation that can be computed from autonomously learned environment maps. Figure 5.4 shows such a learned map of the robot's operating environment and Figure 5.5 a CAD map of the same environment. Both maps contain the main axes Voronoi diagram. Given a map, we can compute its Voronoi diagram, a collection of regions that divide up the map. Each region corresponds to one obstacle, and all the points in one region are closer to the corresponding obstacle than to any other obstacle. The main axes Voronoi skeleton are the set of borderlines between the Voronoi regions that are parallel to the environment's main axes (Figure 5.5). We use main axes Voronoi skeletons instead of ordinary ones because main axes Voronoi skeletons are more robust against inaccuracies of maps that are caused by sensor noise and inaccuracy than their ordinary counterparts.

Based on the Voronoi skeleton we define perceived narrowness as follows. Every point on the Voronoi skeleton has a clearance, its distance to the next obstacle. The narrowness of points is the clearance of the closest point on a Voronoi skeleton. Here, we define that a location is to be perceived as narrow if the distance to the next obstacle is smaller than one robot diameter; but, of course, this value should be learned from experience. Based on these definitions the robot has the following perceptions of narrowness on the path depicted in Figure 5.7(b).

**Diagnosis of Conspicuous Subtraces.** If robots had perfect sensing capabilities and effector control, there were no inertia in the robot's dynamics, and the plans would not be executed by a reactive module that overwrites planned decisions, then the robot would perform exactly as specified in its navigation plans. In practice, however, robots rarely satisfy these criteria and therefore the behavior they exhibit often deviates substantially from what is specified in the control programs. Typically, the effects of the robot's deficiencies depends on the surroundings of the robot. As a rule of thumb: the more challenging the surroundings the larger are the effects of the robot's deficiencies. For example, in wide open areas the deficiencies do not cause substantial problems. In narrow and cluttered surroundings, however, they often cause the robot to stop abruptly, turn in place, etc.

Unfortunately, the navigation behavior is the result of the subtle interplay of many complex factors. These factors include the robot's dynamics, sensing capabilities, surroundings, parameterizations of the control program, etc. It is impossible to provide the robot with a deep model for diagnosing navigation behavior. Instead, we equip the robot with simple heuristic rules such as narrow passages sometimes provoke the robot to navigate with low translational velocity or to stop frequently. Then, the robot is to detect which narrow passages might cause such effects and estimates their extent from experience and by analyzing its behavior.

Let us consider the process of behavior diagnosis more carefully. Upon carrying out a navigation task, RHINO produces a behavior trace like the one shown in Figure 5.6. The robot's position is depicted by a circle where the size of the circle is proportional to the robot's translational speed. Figure 5.7(a) shows those behavior stretches where the robot's translational velocity is low. This is an example of a ***behavior feature subtrace***, a subtrace that represents a feature of the robot's behavior. Figure 5.7(b) shows the behavior stretches that are classified as traversing a narrow passage. This is an ***environment feature subtrace***.

Behavior feature subtraces such as low and high translational speed, turning in place, and frequent stopping often hint at which behavior stretches can

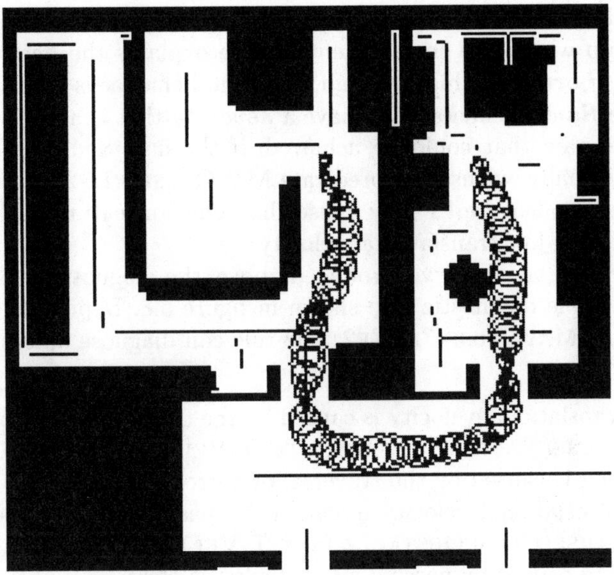

**Fig. 5.6.** Visualization of a behavior trace. The center of the circles denote the robot's position and the size of the circle the current speed of the robot

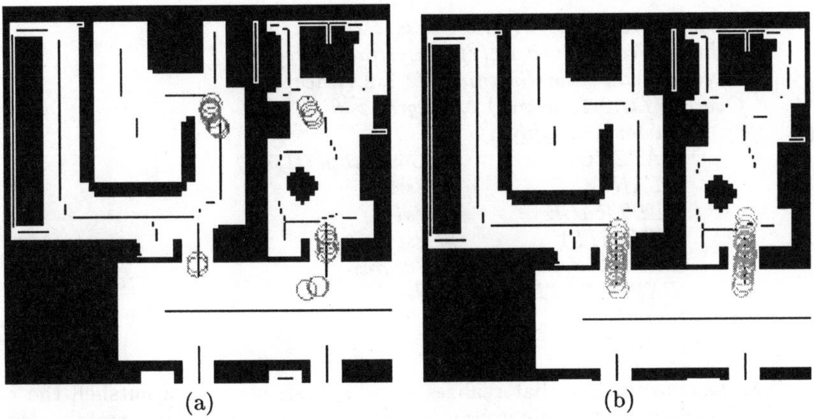

**Fig. 5.7.** Subfigure (a) visualizes the LOW-T-VEL subtraces of the behavior trace depicted in figure 5.6 and subfigure (b) the subtraces that correspond to the traversal of narrow passages

be improved. To infer how the improvements might be achieved, XFRMLEARN first tries to explain a behavior feature subtrace by finding environment feature subtraces that overlap with it. Environment features include being in a narrow passage, in free space, or close to an obstacle. We define the percentage of the behavior subtrace that lies in the environment feature subtrace the degree to which the environment feature explains the behavior. If the percentage of overlap is high enough, then the behavior is considered to be a ***behavior flaw***. Behavior flaws have a ***severity*** that is an estimate of the performance gain that could be achieved, if the flaw would be completely eliminated. Finally, we use the predicate MAYCAUSE(F1,F2) to specify that the environment feature F1 may cause the behavior feature F2 like narrow passages causing low translational velocity.

Based on the vocabulary introduced above, the diagnosis step is realized through a simple diagnostic rule shown in figure 5.8. Depending on the instantiation of MAYCAUSE(?F1,?F2) the rule can diagnose different kinds of flaws:

**D-1** Low translational velocity is caused by the traversal of narrow passages (MAYCAUSE(NARROWPASSAGE,LOW-T-VEL)).
**D-2** Stopping is caused by the traversal of narrow passage.
**D-3** Low translational velocity is caused by passing an obstacle too close (MAYCAUSE(CLOSEOBSTACLE,LOW-T-VEL)).
**D-4** Stopping caused by passing an obstacle too close.
**D-5** High target velocity caused by traversing free space. In this case the robot is driving slower than it should and therefore wastes time.

```
if    BehaviorTrace(?trace) ∧ MayCause(?f2,?f1)
      ∧ BehaviorFeatureSubtrace(?f1,?s1,?trace,?severity)
      ∧ EnvironmentFeatureSubtrace(?f2,?s2,?trace)
      ∧ Overlap(?s1,?s2,?degree) ∧ ?degree > θ
then  infer BehaviorFlaw with
           CATEGORY          = CausedBy(?f1,?f2)
           PLAN              = ?srnp
           SUBTRACE          = ?s1
           SEVERITY          = ?severity
           EXPLAINABILITY    = ?degree
           EXPLANATION       = ?s2
```

**Fig. 5.8.** Diagnostic rule that realizes the diagnosis step. In a nutshell the rule says: if a behavior and an environment feature trace overlap to a certain degree ?DEGREE that exceeds some given threshold and if, according to the domain model, the environment feature may cause the the behavior feature, then the environment feature is said to be a causal explanation for the behavior feature subtrace with degree ?DEGREE

### 5.4.2 The "Revise" Step

The "revise" step uses programming knowledge about how to revise navigation plans and how to parameterize the subsymbolic navigation modules that is encoded in the form of plan transformation rules. In their condition parts transformation rules check their applicability and the promise of success. These factors are used to estimate the expected utility of rule applications. XFRMLEARN selects the rule with a probability proportional to the expected utility and applies it to the plan.

```
transformation rule  traverseNarrowPassageOrthogonally
to eliminate  ?behavior-flaw with
                 CATEGORY          = CausedBy(LowTVel,
                                              NarrowPassage)
                 PLAN              = ?srnp
                 SUBTRACE          = ?subtrace1
                 SEVERITY          = ?severity
                 EXPLAINABILITY    = ?degree
                 EXPLANATION       = ?subtrace2
   if  NavigationStep (?subtrace2,traverseNarrowPassage(?p1,?p2))
        ∧ applicability(traverseNarrowPassageOrthogonally,
                         CausedBy(LowTVel,NarrowPassage),
                         ?applicability),
   then  with  expected-utility = ?severity * ?degree * ?applicability
         insert subplan
            TRAVERSE-NARROW-PASSAGE(?p1,?p2)
              path constraints  { ?p1,?p2 }
              justification       ?behavior-flaw
```

**Fig. 5.9.** The transformational learning rule **R-1**. The rule is applicable if the subtrace *?subtrace* traverses a narrow passage *traverseNarrowPassage(?p1,?p2)*. The application inserts a subplan into the SRNP that constrains the path to cause the robot to traverse a narrow passage orthogonally and with maximal clearance

Figure 5.9 shows a transformation rule indexed by a behavior flaw *CausedBy(NarrowPassage)*. The rule is applicable if the subtrace *?subtrace* traverses a narrow passage *traverseNarrowPassage(?p1,?p2)*. The application inserts a subplan into the SRNP that constrains the path to cause the robot to traverse a narrow passage orthogonally and with maximal clearance.

The main piece of work is done by the condition

NAVIGATIONSTEP(?SUBTRACE,TRAVERSENARROWPASSAGE(?P1,?P2)),

which computes a description of the narrow passage that ?SUBTRACE traverses. ?P1 is then instantiated with the point of the main axes Voronoi skeleton that is next to the first entry in ?SUBTRACE and ?P1 with the one that corresponds to leaving the narrow passage. Using the lines of the main axes Voronoi skeleton to characterize navigation steps achieves more canonical representations of the discretized subtraces and thereby facilitate the identification of identical subtraces.

For the purpose of this discussion, XFRMLEARN provides the following revisions:

**R-1** If the behavior flaw is attributed to the traversal of a narrow passage (**D-1**) then insert a subplan that causes the robot to traverse the passage orthogonally and with maximal clearance.
**R-2** Switch off the sonar sensors while traversing narrow passages if the robot repeatedly stops during the traversal (indexed by **D-2**).
**R-3** Insert an additional path constraint to pass a closeby obstacle with more clearance (indexed by **D-3,D-4**).
**R-4** Increase the target velocity for the traversal of free space where the measured velocity almost reaches the current target velocity (indexed by **D-5**).
**R-5** Insert an additional path constraint to avoid abrupt changes in the robot's heading (indexed by **D-2**).

*Estimating the Expected Utility of Revisions.* Because XFRMLEARN's transformation rules are heuristic, their applicability and the performance gain that can be expected from their application is environment and task-specific. Therefore an important aspect in behavior adaption is to learn the environment- and task-specific expected utility of rules based on experience.

The expected utility of a revision $r$ given a behavior flaw $b$ is computed by

$$EU(r|diagnosed(b)) = P(success(r)|correct(b)) * s(b),$$

where $b$ is a diagnosed behavior flaw, $s(b)$ is the severity of the behavior flaw (the performance gain, if the flaw would be completely eliminated), $P(success(r)|correct(b))$ is the probability that the application of revision $r$ improves the robot's performance significantly given that the diagnosis of $b$ is correct. This formula can be derived from the equation

$$\begin{aligned} &EU(r|diagnosed(b)) \\ &\quad = EU(r|correct(b)) \ * \ P(correct(b)|diagnosed(b)) \end{aligned}$$

In the current version of XFRMLEARN we assume that the diagnoses are correct.[2] In this case, we can transform the equation into

$$\begin{aligned} &EU(r|diagnosed(b)) \\ &\quad = P(success(r)|correct(b)) \ * \ U(success(r)) \\ &\qquad + P(\neg success(r)|correct(b)) \ * \ U(\neg success(r)) \end{aligned}$$

Because XFRMLEARN rejects revisions that do not significantly improve the robot's behavior, the utility of unsuccessful revisions is 0. Furthermore, we assume that $U(success(r))$ is proportional to the severity of the behavior flaw $b$. This leads to the first equation.

For computing the probability

[2] In future versions of XFRMLEARN we would like to learn belief nets for estimating the conditional probabilities that diagnoses are correct.

$$P(success(r)|correct(b)) = P(success(r)|diagnosed(b))$$

(under our assumption above) XFRMLEARN maintains a simple statistic about successful and unsuccessful rule applications. Let $h^+(r,b)$ be the number of times that $r$ successfully revised a SRNP given the behavior flaw $b$ and $h^-(r,b)$ the number of times that the revision was not successful. Then

$$\frac{h^+(r,b)}{h^+(r,b)+h^-(r,b)} \tag{5.1}$$

is a crude measure for $P(success(r)|correct(b))$. $h^+(r,b)$ and $h^-(r,b)$ are initialized to 1.

In general, the first equation overestimates the expected utility of a revision but tends to select, with respect to our practical experience, the same revisions that programmers propose when visually inspecting the behavior traces.

The estimate of the expected utility of the second most promising revision is used as an estimate of how interesting the current navigation task or the SRNP for this task respectively is for future examinations by XFRMLEARN.

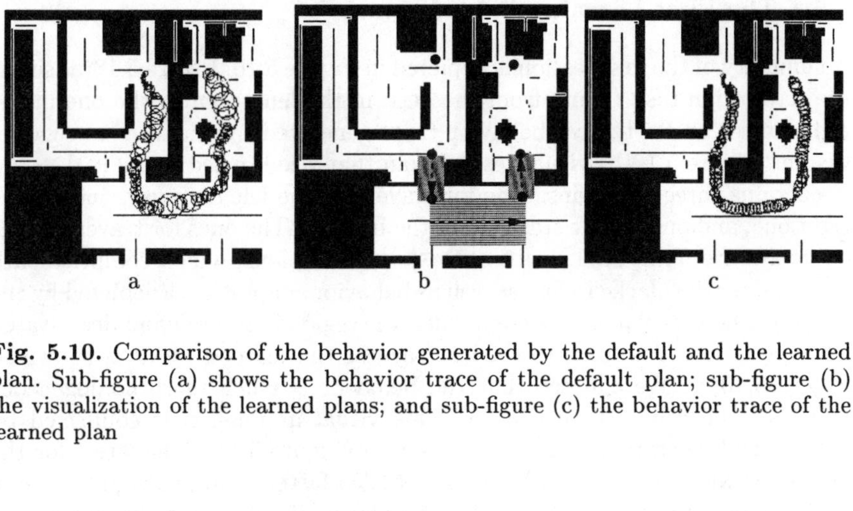

**Fig. 5.10.** Comparison of the behavior generated by the default and the learned plan. Sub-figure (a) shows the behavior trace of the default plan; sub-figure (b) the visualization of the learned plans; and sub-figure (c) the behavior trace of the learned plan

### 5.4.3 The "Test" Step

Because plan transformation rules check their applicability and parameterization with respect to idealized models of the environment, the robot, the perceptual apparatus, and operation of the subsymbolic navigation system, XFRMLEARN cannot guarantee any improvements of the existing plan. Therefore, XFRMLEARN tests the resulting candidate plans against the original plan by repeatedly running the original and the revised plan[3] and measuring the time performance in the local region that is affected by the plan transformation. The new candidate plan is accepted, if based on the experiments there is a 95% confidence that the new plan performs better than the original one. This test only checks whether a revision has locally improved the performance in a statistically significant way. Note that we also use a t-test to show that a sequence of revisions globally improves the robot's performance for a given navigation task with statistical significance.

## 5.5 Experimental Results

To empirically evaluate XFRMLEARN we have performed two long term experiments in which XFRMLEARN has improved the performance of the RHINO navigation system, a state-of-the-art navigation system, for given navigation tasks by up to 44 percent within 6 to 7 hours. Our conclusions are based on these two learning sessions because current robot simulators, which would allow for more extensive experimentation, do not model robot behavior accurately enough to perform realistic learning experiments in simulation.

### 5.5.1 The First Learning Experiment

A summary of the first session is depicted in Figure 5.10. Figure 5.10(a) shows the navigation task (going from the desk in the left room to the one in the right office) and a typical behavior trace generated by the MDP navigation system. Figure 5.10(b) visualizes the plan that was learned by XFRMLEARN. It contains three sub-plans. One for traversing the left doorway, one for the right one, and one for the traversal of the hallway. The ones for traversing the doorways are TRAVERSENARROWPASSAGE sub-plans, which comprise path constraints (the black circles) as well as behavior adaptations (depicted by the region). The sub-plan is activated when the region is entered and deactivated when it is left. A typical behavior trace of the learned SRNP is shown in Figure 5.10(c). We can see that the behavior is much more homogeneous and that the robot travels faster. This visual impression is confirmed by the descriptive statistics of the comparison (Figure 5.11). The t-test for the learned SRNP being at least 24 seconds (21%) faster returns a significance of 0.956. A bootstrap test returns the probability of 0.956 that the variance of the performance has been reduced.

In the session XFRMLEARN has proposed and tested five plan revisions, three of them were successful. The unsuccessful ones proposed that the low translational at the beginning of the behavior trace was caused by the robot passing an obstacle too closely. Indeed it was caused by the robot not facing the direction it was heading to and the inertia of the robot while accelerating.

[3] In our experiments we run both plans 7 times.

| | ORIGINAL PLAN | LEARNED PLAN |
|---|---|---|
| Minimum duration | 85.2 | 67.5 |
| Maximum duration | 156.7 | 92.6 |
| Range | 71.5 | 25.0 |
| Median | 111.0 | 76.5 |
| Mean | 114.1 | 78.2 |
| Variance | 445.2 | 56.5 |
| Deviation | 21.1 | 7.5 |

**Fig. 5.11.** The table compares the descriptive statistics for the duration of the navigation task using the original and the learned plan. We can see that the robot performs the navigation task faster and with less variance using the learned plan

The second unsuccessful revision proposed to go faster in the open area of the left office. The performance gain obtained by this revision was negligible. Figure 5.12 summarizes the proposed revisions and whether they have been successful or not.

| | Rule | Applied To | Accepted? |
|---|---|---|---|
| 1 | R-1/2 | door right | yes |
| 2 | R-1/2 | door left | yes |
| 3 | R-4 | hallway | yes |
| 4 | R-4 | room left | no |
| 5 | R-3 | room left | no |

**Fig. 5.12.** Summary of the revisions performed by XFRMLEARN in the learning experiment. The rule names refer to the rules that are listed on page 135. The first column specifies the learning cycle, the second one the name of the applied rule, the third specifies the region that the revision applies to, and the last column states whether or not the revision has been accepted in the test phase

### 5.5.2 The Second Learning Experiment

Fig 5.13 summarizes the results from the second learning session: the learned SRNP (left) and the descriptive statistics (right). The average time needed for performing a navigation task has been reduced by about 95.57 seconds (44%). The t-test for the revised plan being at least 39 seconds (18%) faster returns a significance of 0.952. A bootstrap test returns the probability of 0.857 that the variance of the performance has been reduced. The successful revisions of the second learning session are visualized in Figure 5.14.

### 5.5.3 Discussion of the Experiments

In this section we discuss three questions that we feel critical for the evaluation of the XFRMLEARN approach. These questions are: (1) Are the time

| | ORIGINAL PLAN | LEARNED PLAN |
|---|---|---|
| Minimum of durations | 145.3 | 86.1 |
| Maximum of durations | 487.6 | 184.5 |
| Range | 342.2 | 98.4 |
| Median | 181.5 | 108.6 |
| Mean | 212.6 | 119.0 |
| Variance | 9623.1 | 948.0 |
| Deviation | 98.0 | 30.7 |

**Fig. 5.13.** Descriptive statistics for the second learning session. Again the learned navigation plan performs the navigation task faster and with less variance

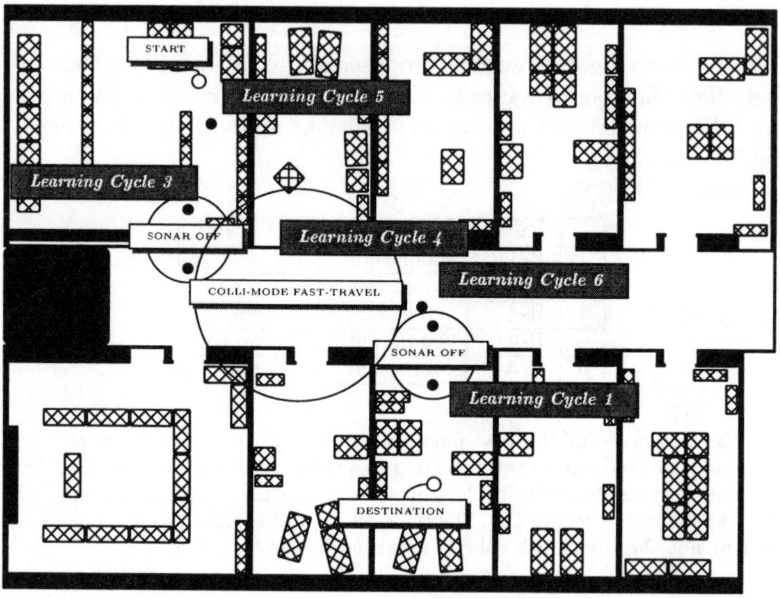


**Fig. 5.14.** Visualization of the second learning experiment. The picture shows the transformations that were considered in the learning cycles of the learning session. The small black circles depict additional path constraints. The big transparent circles represent travel mode adaptations and the regions in which they are active

resources needed for the learning sessions justified by the performance gains obtained through them? (2) Is the performance gain not only due to a small set of heuristic and highly tailored rules? (3) Is it really necessary to make such extensive tests before making a learning step? Or, doesn't reinforcement learning do the same in a simpler and more effective way? We will answer these questions in the subsequent sections.

**Performance Gains.** We consider the performance gains that XFRMLEARN has achieved in the learning sessions described in Section 5.5 as being very impressive, for several reasons. First, XFRMLEARN has only used observable symptoms such as driving slowly or fast. It did not have access to the execution states of the collision avoidance module which would have provided more informative and reliable evidence for the diagnosis of behavior flaws.[4] Second, the transformation rules are quite general and applicable to many indoor environments. They are also not specifically tailored for this particular system. Third, we have applied XFRMLEARN to a low-level navigation system which is well tuned and has been developed in the same environment. We expect the performance gains for insufficiently configured navigation systems to be much higher.

**Generality.** An important issue that still needs to be discussed is that of generality. In this paper we have only addressed the navigation problem in indoor environments. We believe that our approach carries over to other tasks in which Markov decision problems are solved and executed by a reactive execution component. MDPs will try to compute policies that generate the fast trajectories through the state space to the goal state. Where the real trajectory significantly deviates from the optimal ones there are often reasons that force the reactive execution component to correct the precomputed trajectory and which have been idealized away in the formulation of the Markov decision problem. To apply our approach to other problem areas is on our agenda for future research.

An important design criterion is the generality of the diagnosis and transformation rules. General rules assume little knowledge about the low-level navigation system and can therefore be adapted with little effort to other environments, robots, and tasks. The rules described in the previous sections are general: switching off sonar sensors in doorways (if it is safe), introducing additional intermediate points to constrain the robot's path, circumnavigate obstacles with greater clearance, and increase and decrease target velocity.

Another important design criterion is how much to generalize the experiences. In this paper we have considered the problem of learning instances of navigation tasks. XFRMLEARN applies revisions to SRNP instances and tests the success by comparing SRNP instances. This way XFRMLEARN avoids the problem of overgeneralization but also cannot exploit the advantages of generalization.

To demonstrate the capability of XFRMLEARN in improving a robot's performance we have used experiments that allow for an easy measurement of the performance gain. However, our goal is to learn universal SRNPs, that is SRNPs that improves on average the behavior of the robot for arbitrary navigation tasks.

**Dealing with the Variance in Robot Behavior.** In our view, one of the main hindrances in the application of machine learning techniques to autonomous robot control is the high variance in robot behavior that is inherent

[4] We plan to make these execution states, such as "current path blocked" or "no admissible trajectory", explicit and to exploit them in future versions.

in most systems that control physical robots in complex environments. This variance makes it very difficult to decide whether one control strategy is better than another one. Therefore it is a main idea of XFRMLEARN to explicitly take this variance into account.

To find a good parameterization of the control system we perform a search in the space of possible parameterizations. As this search space is continuous we consider a greedy search strategy as promising. Because accepting worse parameterizations as improvements would cause a hill-climbing learning algorithm to search into the wrong direction, the risk of making such wrong decisions should be minimized. Therefore XFRMLEARN uses a stringent test: a new candidate parameterization is accepted only if it increases the performance significantly (in a statistical sense).

In our learning experiments we have to deal with various sources of variance. For example, the time for traversing a doorway depends on whether or not sonar cross talk occurs in a particular trial, how close a particular path passes the door frame, the amount of rotation upon entering the doorway, etc.

| DAY 1 | | DAY 2 | |
|---|---|---|---|
| ORIGINAL | LEARNED | ORIGINAL | LEARNED |
| 127.3 | 107.8 | 265.9 | 184.5 |
| 172.2 | 112.9 | 148.5 | 97.9 |
| 131.4 | 110.8 | 487.6 | 108.6 |
| 169.2 | 108.0 | 195.3 | 138.3 |
| 122.0 | 106.8 | 167.4 | 95.6 |
| 220.2 | 106.8 | 145.3 | 93.6 |

**Fig. 5.15.** The table shows the variance of RHINO's performance controlled by the same SRNP on two subsequent days. While the durations vary drastically, the relative results of the comparison are qualitatively similar. The data show the importance of statistical tests for the acceptance of candidate plans

Given such high variances, passing the significance test requires a large number of time consuming experiments. To reduce the number of experiments that are needed we employed the following techniques. First, to compare the original SRNP with the revised one, XFRMLEARN first determines the region in which the behavior of the robot might be affected by the revision. As test region we take the (1 meter) envelope around the region that is directly effected by the transformation. This ensures that every side effect the transformation possibly could have is taken into account when comparing the two SRNPs. Second, to reduce the effect of variance producing factors that cannot be influenced two SRNPs are tested alternatingly. Third, we use macro revision rules, such as transformation rule **R-6** that perform multiple revisions to eliminate a combination of behavior flaws that occurred in the same region. Our experiments have shown that in many cases neither rule **R-2** nor **R-1** were capable of improving the traversals of doorways significantly whereas rule **R-6** could do so with extremely high significance. A lesson that we draw from our experiences is that a learning system for tuning robot behavior should not only apply SRNP revisions to improve the robot's performance but also ones that reduce the variance without decreasing the robot's performance.

Generating one data pair, a behavior trace for the original and the revised plan takes, about eight minutes. Thus the net execution time for the learning session is about 280 minutes, almost five hours. This price must be paid for making low risk decisions about which SRNP produces better behavior. The variance in the physical performance prevents to rely on fewer experiments.

Given that we use such a stringent criterion for the acceptance of new candidate SRNP the time needed for tests is reasonable. To test one SRNP revision takes about 60 minutes and a complete learning session for one navigation task takes about 6-8 hours. These time requirements, however, can be drastically reduced, probably by a factor 3-4, if (1) a navigation task and its inverse task are learned together and (2) using local tests several hypotheses can be tested simultaneously. For example, two revisions for traversing different doors could be tested simultaneously without having to consider interferences between the two revisions. These optimizations would imply that an office delivery robot could autonomously adapt itself to its working environment within a weekend. Further reductions can be obtained by adequately generalizing learning results. We decided not to try these optimizations during XFRMLEARN's development phase.

## 5.6 Related Work on Learning Robot Plans

XFRMLEARN can be viewed as an extension of the transformational planning system XFRM (McDermott, 1992b). Using a library of modular default plans, XFRM computes a plan for a given compound task and improves the plan during its execution based on projections of its execution in different situations. XFRMLEARN is a step towards learning default plan libraries. XFRMLEARN differs from other approaches that build up plan libraries by plan compilation/explanation-based learning (Laird, Rosenbloom, and Newell, 1986; Mitchell, 1990) in that it does not require the existence of symbolic action representations but (partly) generates the symbolic plans itself.

A number of researchers consider navigation as a (Partially Observable) Markov decision problem (MDPs) (Thrun et al., 1998; Kaelbling, Cassandra, and Kurien, 1996) and compute *policies*, mappings from discretized robot poses into fine-grained navigation actions, that cause the robot to reach its destination quickly and reliably. This approach has been successfully applied to the problem of robot navigation. Though MDP path planners produce a behavior that is robust and optimized for average performance, they are not suitable for handling exceptional situations. We have also discussed in the introductory section of this chapter several advantages of hierarchically structured plans for high-level control of complex continuous navigation behavior.

In our research we combine (PO)MDP planning and transformational planning of reactive behavior. MDP path planning is used for generating robust navigation behavior optimized for average performance and transformational planning for predicting and forestalling behavior flaws in contingent situations. XFRMLEARN bridges the gap between both approaches: it learns symbolic plans from executing MDP policies and it can be used to decompose MDP problems and specify the dimensions and the grain size of the state space for the subproblems (Moore, 1994). Our approach has been inspired by Sussman's thesis that failures (bugs) are valuable resources for improving the performance of problem-solving agents (Sussman, 1977).

Our work is also related to the problem of learning actions, action models or macro operators for planning systems. Oates, Schmill, and Cohen (2000) recently proposed a technique for learning the pre- and postconditions of robot actions. Sutton, Precup, and Singh (1998) have proposed an approach for learning macro operators in the decision-theoretic framework using reinforcement learning techniques. The IMPROVE algorithm (Leash, Martin, and Allen, 1998) runs data mining techniques on simulations of large, probabilistic plans in order to detect defects of the plan and applies plan adaptations to avoid these effects.

Mahadevan (1996) classifies existing approaches to robot learning into reinforcement, evolutionary, explanation-based, and inductive concept learning. This section discusses how these four approaches can be applied to learn parameterizations for complex robot control systems.

Both, reinforcement learning (Sutton and Barto, 1998) and evolutionary learning (Koza, 1992) have been applied to similar problems. Floreano and Mondada (1998), for example, use evolutionary learning for learning the parameterization of a simple robot control system. Santamaria and Ram (1997) describe an approach for learning a so-called adaptive policy, a mapping from perceived situations to the continuous space of possible configurations of a purely reactive navigation system. They consider the problem as a reinforcement learning problem and solve it using a hybrid learning algorithm that combines techniques from case-based learning and reinforcement learning. Both kinds of approaches are very elegant and do not require (sophisticated) models. It would be very interesting and instructive to test these general and model-free approaches on hand-tuned and sophisticated controllers (such as the RHINO system) where the parameterizations that improve the performance are very sparse in the space of possible parameterizations. For the adaptation of sophisticated controllers the use of models is a valuable resource to find parameterizations that are capable of improving the behavior. In addition, we only know of applications of these learning techniques to purely reactive controllers. It remains to be investigated whether pure mappings from perceptions to parameterizations are capable of producing satisfying global behavior. For example, in the context of service robotics certain parameter configurations should be associated with environment regions rather than perceived situations in order to achieve more coherent and predictable navigation behavior.

Goel et al. (1997) introduce a framework that is, in some aspects, similar to ours: A model-based method monitors the behavior generated by a reactive robot control system, detects failures in the form of behavioral cycles, analyzes the processing trace, identifies potential transformations, and modifies the reactive controller. However, the method is used to reconfigure a reactive control system when it is trapped in a cyclic behavior due to the lack of a global perspective on the task it has to perform. They only perform revisions to resolve conflicts between different reactive behaviors. In addition, their revisions are specific to particular problem episodes and cannot be reused for other episodes. Therefore they rather solve a planning problem using transformational planning techniques.

Dependency interpretation (Howe and Cohen, 1992) uses statistical dependency detection to identify patterns of behavior in execution traces in order to detect possible causes of plan failures. Haigh and Veloso propose a method for learning the situation-specific costs of given actions (Haigh and Veloso, 1998) whereas our approach aims at learning and improving the actions themselves. The main differences to our approach include that our approach also learns the symbolic actions and adapts and improves continuous control processes.

Though we have reported that XfrmLearn improves the performance of the mobile robot Rhino significantly and substantially this was not our goal in it's on right. In that our approach has to be delimited from other work that applies well-known learning techniques to the tuning of the performance of a mobile robot(Sutton and Barto, 1998; Floreano and Mondada, 1998; Santamaria and Ram, 1997).

Explanation-based learning techniques (Mitchell, 1990) cannot be easily applied to our learning task because they require complete and correct models of the robot and its interaction with its environment, which can rarely be provided. Explanation-based neural net learning (Mitchell and Thrun, 1993; Towell and Shavlik, 1989) does not depend on such correct and complete domain theories, but does not seem to be applicable to our problem. Our approach can also be regarded as a form of explanation-based learning that can use knowledge that is not required to be complete and correct.

## 5.7 Discussion

In our view, the performance gains over MDP navigation policies indicate that the syntactic structure of the plans represents an advantageous discretization of the continuous navigation behavior.

We have described XFRMLEARN, a system that learns SRNPs, symbolic behavior specifications that (a) improve the navigation behavior of an autonomous mobile robot generated by executing MDP navigation policies, (b) make the navigation behavior more predictable, and (c) are structured and transparent so that high-level controllers can exploit them for demanding applications such as office delivery.

XFRMLEARN is capable of learning compact and modular SRNPs that mirror the relevant temporal and spatial structures in the continuous navigation behavior because it starts with default plans that produce flexible behavior optimized for average performance, identifies subtasks, stretches of behavior that look as if they could be improved, and adds subtask specific sub-plans only if the sub-plans can improve the navigation behavior significantly.

The learning method builds a synthesis among various subfields of AI: computing optimal actions in stochastic domains, symbolic action planning, learning and skill acquisition, and the integration of symbolic and subsymbolic approaches to autonomous robot control. Our approach also takes a particular view on the integration of symbolic and subsymbolic control processes, in particular MDPs. In our view symbolic representations are resources that allow for more economical reasoning. The representational power of symbolic approaches can enable robot controllers to better deal with complex and changing environments and achieve changing sets of interacting jobs. This is achieved by making more information explicit and representing behavior specifications symbolically, transparently, and modularly. In our approach, (PO)MDPs are viewed as a way to ground symbolic representations.

# 6. Plan-Based Robotic Agents

This chapter describes three autonomous robotic agents that we have realized and controlled with structured reactive controllers. The first one, described in section 6.1, is the robot office courier that we have already seen in chapter 1. Section 6.2 describes a plan-based robotic museums tour-guide and a long-term experiment that demonstrates the reliability of SRCs. A robot party butler is briefly sketched in section 6.3. The chapter concludes with several demonstrations of the integrated mechanisms for image processing, communication, active localization, and object search (Section 6.4).

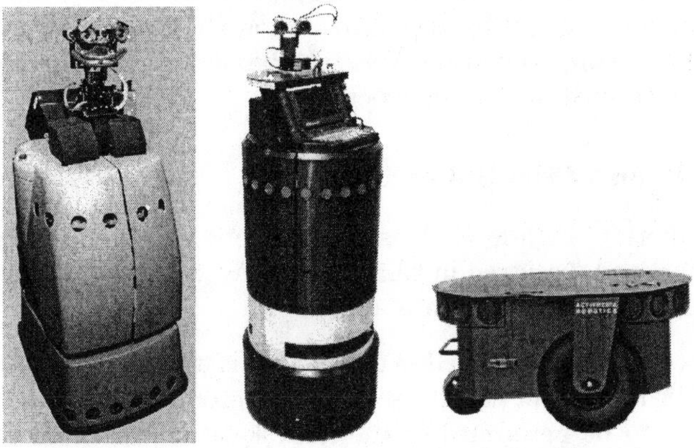

(a) (b) (c)

**Fig. 6.1.** The mobile robots MINERVA (a), RHINO (b), and JAMES that are used in the following experiments

## 6.1 A Robot Office Courier

We have implemented an SRC to control RHINO (Figure 6.1(b)) and carry out delivery tasks in an office environment. The robot courier is not allowed to carry letters of the same color at the same time. The SRC is designed to control the robot over extended periods of time and carry out schedules of multiple requests, which can be changed at any time. The task of the plan-based controller is to flexibly and reliably carry out planned delivery tours, where the delivery tours are specified as structured reactive plans. The high-level controller manages its intended plan for the delivery tour according to the changes with respect to the requests it is to satisfied and its beliefs about the state of the environment.

To perform its tasks fast RHINO schedules the pick-up and delivery actions as to minimize execution time. To ensure that its schedules will work, RHINO makes sure that letters are picked up before delivered and predicts how the world changes and what the effects of these changes on its scheduled activities might be. Constraints on states that schedules must satisfy are that RHINO does not carry two letters with the same color and that it meets the deadlines.

The belief state of the robot is maintained with respect to the electronic mails that the robot receives, the sensor data that it obtains, and using rules for updating the beliefs about dynamic states as time passes. Electronic mails often contain information about the state of the environment. A typical rule that describes the dynamics of the environment is the following one: "if office A-111 is closed it typically stays closed for about fifteen minutes" specifies that if the robot has not received any evidence about the door of room A-111 for fifteen minutes, the probability distribution for the door state is reset to the a priori probability. Haigh and Veloso (1998) describe a learning technique that can acquire such rules from experience.

### 6.1.1 The Plans of the Robot Courier

The most important high-level plans of the robotic office courier have already been described and discussed in Chapter 3.3. We will only recapitulate their purpose and effects. The plans are:

at location $\langle x,y \rangle$ $p$, which specifies that plan $p$ is to be performed at location $\langle x, y \rangle$. The purpose of the plan schema is to group together a set of actions that have to be performed at the same location, make the location at which they are to be performed explicit, and give the sub-plan a name tag so that plan management operations can refer to and manipulate it easily.

achieve *ObjLoc(obj,*$\langle x,y \rangle$*)*, which specifies how object *obj* is to be delivered to location $\langle x,y \rangle$. Essentially, the task is carried out by performing a pickup action at the location where the *obj* is initially and a putdown action at the destination $\langle x,y \rangle$ of the object.

Recall that the plans are designed to be ***general***, ***embeddable***, ***restartable***, and ***location transparent*** (see Chapter 3.3).

The structure of the top-level plan is sketched in Figure 6.2. The plan contains stabilizers (line 1 to 2). Stabilizers are sub-plans that, when triggered, reachieve some state that is necessary for the successful operation of the robot. In our case, it is necessary to relocalize the robot whenever it loses track of its position, and to reload the batteries whenever they are getting empty. The outer plan adaptors in line 3 and 4 are plan adaptors that are always active. They perform tasks like integrating sub-plans for new user requests, removing plans for user requests that have been completed or that the robot does not believe to be successfully completeable. The inner part of the plan (line 5 - 10) is the part of the plan that the controller reasons about and manipulates. This part also contains stabilizers and plan adaptors but those are situation specific and inserted by the outer plan management operations. The inner plan adaptors handle situations like how to deal with a specific opportunity. Finally, the body of the plan (line 8 - 10) contains the individual delivery plans that contain the at location sub-plans and an order on these at location sub-plans that specifies the delivery tour.

```
top-level-plan OFFICECOURIER( )
 1  with stabilizer relocalize when necessary
 2                  reload battery when necessary
 3  do with plan adaptor manage the plan with respect to the user's requests
 4                       and the robot's utility model
 5    do named subplan main − plan
 6    do with stabilizer situation specific stabilizers
 7      do with plan adaptor situation specific plan adaptors
 8        do partial order set of delivery plans
 9                         containing at location sub-plans
10          :order ordering constraints on at location sub-plans
```

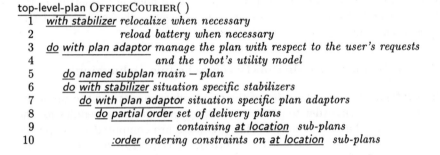

**Fig. 6.2.** Schematic structure of the top-level controller of the robot office courier

The plan adaptors essentially modify sub-plans and restart the sub-plans from anew if they have already been partially executed. Because the plans are restartable the revised sub-plan continues with the execution where the original sub-plan has stopped.

### 6.1.2 Plan Adaptors of the Robot Courier

The SRC of the robot office courier uses static plan adaptors for the following kinds of belief changes:

- asynchronously arriving new requests and revisions and deletions of existing ones;
- doors that are expected to be open but detected to be closed and vice versa;
- colors of letters to be delivered that turn out to be different from their expected colors; and
- unexpected delays in delivery tours.

The plan adaptors that deal with user request changes use revision rules for the insertion of sub-plans for accomplishing the new requests, the deletion of sub-plans for requests that have been completed or the ones that are not believed to be completeable. If requests are revised the sub-plans for the original requests are deleted and the ones for the revised request are inserted. The plan adaptors that deal with closed doors turn sub-plans that cannot be completed because of the closed door into opportunistic sub-plans that are triggered by the corresponding door becoming open. The plan adaptors for dealing with doors that have become open are associated with a tactical and a strategic plan revision rule. The tactical revision rules computes a schedule fast but does not take predictable schedule flaws into consideration. The strategic revisions are slower but typically produce better schedules. Finally, the robot courier uses rescheduling revisions that schedule sub-plans with tight deadlines earlier. Revision rules applied by the robot courier also include one that adds ordering constraints to avoid holding two letters of the same color and one that inserts an email action asking the sender of a letter to use a particular envelope into the plan.

### 6.1.3 Probabilistic Prediction-Based Schedule Debugging

The strategic plan revision rules of the robot office courier use the predicate GOODSCHEDULE in order to generate working and efficient schedules for the delivery tours. The extensions of this predicate are computed by a specific inference technique that we call ***Probabilistic Prediction-based Schedule Debugging*** (PPSD). PPSD is a resource-adaptive algorithm, that generates an initial heuristic schedule under simplified conditions fast and improves on the schedule given more computational resources.

```
algorithm ppsd(PLAN, BS, N)
  ⟨STEPS,O⟩ ← PLAN
  loop
      O' ← GENERATE-SCHEDULE(⟨STEPS,O⟩)
      ES ← RANDOMLY-PROJECT(⟨STEPS,O'⟩, BS, N)
      FLAWS ← DETECT-SCHEDULE-FLAWS(ES,K)
      RULE ← CHOOSE(RULES(CHOOSE(FLAWS)))
      ⟨STEPS,O⟩ ← APPLY(RULE)
    until FLAWS = {}
```

**Fig. 6.3.** Computational structure of the prediction-based schedule debugger

PPSD solves the following computational task: Given a probabilistic belief state, a set of jobs, and a delivery plan with a partial order on the *at location* subplans, find a total ordering on the *at location* subplans, that accomplishes the given jobs flawless and as fast as possible. In the robot office courier application the probabilistic belief state comprises probability distributions over which envelopes are located on which desks and the current state of doors. They are encoded in the form of probability distributions over the values of predefined random variables (e.g., *P(two-yellow-letters-on-desk-3) = 0.7*).

The computational structure of prediction-based schedule debugging (figure 6.3) is a cycle that consists of the following steps: First, call a heuristic ***schedule generator*** to produce a candidate schedule O′. The schedule generator uses heuristics and domain-specific methods for quickly computing schedules that minimize travel time. It does not take into account the state changes of the robot and the world that occur as the activities get executed. Second, randomly sample N projected execution scenarios ES of the plan ⟨STEPS,O′⟩ constrained by a set of policies with respect to the belief state BS. Third, apply the flaw detection module to the execution scenarios ES and collect all flaws FLAWS that occur at least K times in the N scenarios. Next, RULE is set to a revision rule that is applicable to one of the flaws in FLAWS. The new plan ⟨STEPS,O⟩ results from the application of RULE to the current plan. The loop is iterated until no probable flaw in the schedule is left. The following pseudo code sketches the main computational steps of the PPSD algorithm.

Before we will describe the individual steps of the algorithm in greater detail, we want to make two important remarks. First, since the schedules computed by PPSD are installed using the revision rules that we have discussed in Chapter 3.4, the delivery plans are location transparent. Second, there is, in general, no guarantee that eliminating the cause for one flaw won't introduce new flaws or that debugging will eventually reduce the number of flaws (Simmons, 1992).

**The Schedule Generator** sorts navigation tasks (counter)clockwise to get a candidate schedule. After this initial sort the scheduler iteratively eliminates and collects all steps $s$ such that $s$ has to occur after $s'$ with respect to the required ordering constraints but occurs before $s'$ in the candidate schedule. Then the steps $s$ are iteratively inserted into the candidate schedule such that the ordering constraints are satisfied and the cost of insertion is minimal. While this greedy algorithm is very simple and fast it tends to produce fast schedules because the benign structure of the environment.

**The Flaw Detection Module** is realized through plan steps that generate "fail" events. Thus, to detect schedule flaws an XFRM-ML query has to scan projected scenarios for the occurrence of failure events. For example, to detect deadline violations we run a monitoring process that sleeps until the deadline passes concurrently with the scheduled activity. When it wakes up it checks whether the corresponding request has been completed. If not a deadline violation failure event is generated.

The flaw detector classifies a flaw as to be eliminated if the probability of the flaw with respect to the agent's belief state exceeds a given threshold $\theta$.

PPSD classifies a flaw as hallucinated if the probability of the flaw with respect to the agent's belief state is smaller than $\tau$. The details of designing resource-adaptive flaw detectors and their reliability using the randomly sampling projection mechanism can be found in section 4.4.

A number of approaches have been applied to activity scheduling in autonomous agent control. McDermott (1992b) has developed a prediction-based scheduler for location specific plan steps, which can probabilistically guess locations at which plan steps are executed if the locations are not specified explicitly. Our *at location* subplan is a variant of the one that has been proposed in McDermott's paper. The main contribution of our approach is that McDermott's approach has been applied to a simulated agent in a grid-based world whereas ours controls a physical autonomous robot. Pell et al. (1997) (re)schedule the activities of an autonomous spacecraft. While their system must generate working schedules we are interested in good schedules with respect to a given utility/cost function. In addition, scheduling activities of autonomous service robots typically requires faster and resource adaptive scheduling methods. Alami et al. (1998) have applied scheduling to coordinate multiple robots, an issue we haven't addressed yet. Our scheduling approach differs from the one proposed by McVey et al. (1997) in that theirs generates real-time guaranteed control plans while our scheduler optimizes with respect to a user defined objective function.

Probabilistic prediction-based schedule debugging is a planning approach rooted in the tradition of transformational planners, like HACKER (Sussman, 1977) and GTD (Simmons, 1992) that diagnose "bugs" or, in our case, plan failures, in order to revise plans appropriately. In spirit, the control strategy of PPSD is very similar to the Generate/Test/Debug strategy proposed by Simmons (1992). Our approach, like the XFRM system (McDermott, 1992b), differs from other transformational planners in that it tries to debug a simultaneously executed plan instead of constructing a correct plan. Also, we reason about full-fledged robot plans and are able to diagnose a larger variety of failures.

### 6.1.4 Demonstrations and Experiments

We have performed an experiment consisting of two steps in order to show that PPSD-based scheduling can improve the performance of the robot office courier. In the first step, we evaluated the robot courier using PPSD in a series of 12 randomly generated scenarios with about six delivery jobs most of them added asynchronously. In these scenarios no plan revisions were necessary. Each scenario was tested in five runs taking about twenty to twentyfive minutes. As expected PPSD-based scheduling did on average not worse than situation-based scheduling methods. In the second step we made up five scenarios in which the computation of good schedules required foresight. In those scenarios PPSD outperformed situation-based scheduling by about eight percent.

In the remainder of this Section we will describe two example runs performed by the robot office courier. In the first one the robot courier only employs situation-specific execution-time rescheduling. In the second one it employs PPSD-based runtime rescheduling.

**Performing Office Delivery Tasks with Situation-Specific Plan Management.** Consider the following experiment that is carried out by RHINO using an SRC and the scheduling capabilities described in (Beetz and Bennewitz, 1998). In the beginning, RHINO carries out no primary activities. Its outermost plan adaptor ensures that new requests are received and processed.

with plan adaptor *integrate revisions of requests* **(P-1)**
with plan adaptor *replan/reschedule when necessary* **(P-2)**
PRIMARY ACTIVITIES

RHINO receives two requests shown in 6.4(a). In the first one RHINO is asked to get a book from Wolfram's desk in A-111 to Jan's desk in A-113. The second one requires RHINO to deliver a letter from "nobody's" desk in A-117 to the library (room A-110). Whenever the jobs change RHINO computes an appropriate schedule very fast. Upon receiving the two requests the plan adaptor **P-1** puts plans for the requests into the plan. The insertion of the requests triggers the scheduler of the plan adaptor **P-2** that orders the AT-LOCATION tasks. The scheduling plan adaptor also adds an additional plan adaptor **P-3** that monitors the assumptions underlying the schedule, that is that the rooms A-110, A-111, A-113, and A-120 are open.

with plan adaptor *integrate revisions of requests* **(P-1)**
with plan adaptor *replan/reschedule when necessary* **(P-2)**
with plan adaptor *reschedule when A-110, A-111, A-113, or A-120 are closed* **(P-3)**
PRIMARY ACTIVITIES
Plan for: *Deliver the book ...*
NAMED P-1 AT-LOCATION $\langle x_1, y_1 \rangle$ pick up book
NAMED P-2 AT-LOCATION $\langle x_2, y_2 \rangle$ deliver book
Plan for: *Deliver the letter ...*
NAMED P-3 AT-LOCATION $\langle x_3, y_3 \rangle$ pick up letter
NAMED P-4 AT-LOCATION $\langle x_4, y_4 \rangle$ deliver letter
:ORDER P-1 $<$ P-2 $<$ P-3 $<$ P-4

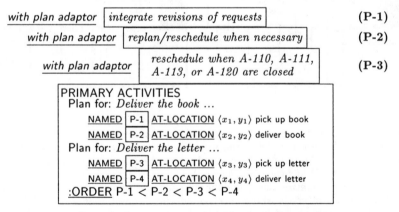

The order of the delivery steps are that RHINO will first go to A-111 to pick up the book and deliver it in A-113. After the delivery of the book it will pick up the letter and deliver it in the library. This is the schedule that implies the shortest path while making sure that the objects are picked up before they are delivered.

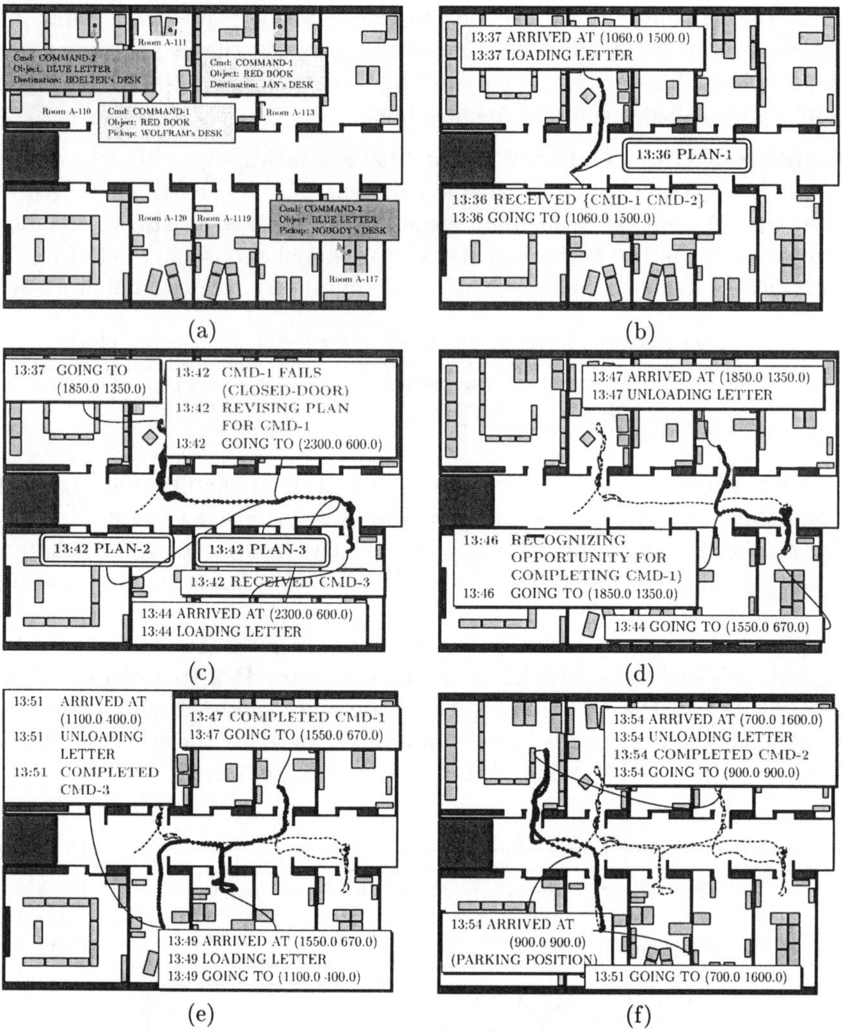


**Fig. 6.4.** Execution trace for the experiment on situation-specifice runtime plan adaptation. Subfigure (a) shows the requests the robot courier is to satisfy. The robot goes first into room A-111 and loads the red book and intends to deliver the red book next (b). While the robot leaves A-111 room A-113 is closed. On its way to A-113 the robot perceives that the door is closed and revises its plan and executes PLAN-2. Shortly after the robot receives another request that yields PLAN-3. It arrives at room A-117 and loads the blue letter (c). On its way to to pick up the item mentioned in the last request it recognizes room A-117 as being now open and delivers the red book as an opportunity (d). It then continues with satisfying the last request (e) and completes its tour in subfigure (f)

Upon receiving the jobs RHINO starts navigating to Wolfram's desk where the book is loaded (Figure 6.4(b)). After RHINO has picked up the book and left room A-111, it notices that room A-113 is closed (Figure 6.4(c)). Because RHINO cannot complete the delivery of the book the corresponding request fails. This failure triggers the plan adaptor **P-3**. Because room A-113 is usually closed it transforms the completion of the delivery into an opportunity. Thus, as soon as RHINO notices room A-113 to be open it interrupts its current mission, completes the delivery of the book, and continues with the remaining missions after the book has been successfully delivered.

*with plan adaptor* | *integrate revisions of requests* | **(P-1)**
*with plan adaptor* | *replan/reschedule when necessary* | **(P-2)**
*with plan adaptor* | *reschedule when A-110, A-111, or A-113 are closed* | **(P-3)**

PRIMARY ACTIVITIES
*with constraining plan* WAIT-FOR OPEN?(A-120)
Plan for: Deliver the book
Plan for: Deliver the letter

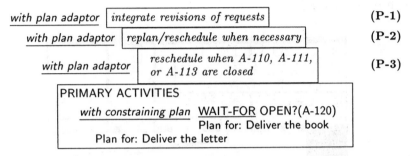

The new scheduled activity asks RHINO to pick up the letter in A-117 and deliver it in the library. In addition, RHINO receives a third request ("deliver the letter from Oleg's desk in A-119 to Michael's desk in A-120") which is integrated into the current schedule while the schedule is executed. The new order is to get first the letter from A-117 and then the letter from A-119 and to deliver the first letter in A-120 and and then the second in the library. Thus, RHINO goes to A-117 to get the letter (Figure 6.4(d)) and continues to navigate to room A-119. As it passes room A-113 on its way to A-119 it notices that the door is now open and takes the opportunity to complete the first request (Figure 6.4(d)). After that it completes the remaining steps as planned and goes back to its parking position where it waits for new requests (Figure 6.4(e-f).

### 6.1.5 Prediction-Based Plan Management

The next experiment shows an SRC predicting and forestalling execution failures while performing its jobs. Consider the following situation in the environment pictured in Figure 1.2. A robot office courier is to deliver a letter in a yellow envelope from room A-111 to A-117 (*cmd-1*) and another letter for which the envelope's color is not known from A-113 to A-120 (*cmd-2*). The robot has already tried to accomplish *cmd-2* but because it recognized room A-113 as closed (using its range sensors) it revised its intended course of action into achieving *cmd-2* opportunistically. That is, if it later detects that A-113 is open it will interrupt its current activity and reconsider its intended course of action under the premise that the steps for accomplishing *cmd-2* are executable.

To perform its tasks fast the robot schedules the pick-up and deliver actions to minimize execution time and assure that letters are picked up before delivered. To ensure that the schedules will work, the robot has to take into account how the state of the robot and the world changes as the robot carries out its scheduled activities. Aspects of states (state variables) that the robot has to consider when scheduling its activities are locations of the letters. Constraints on the state variables that schedules have to satisfy are that they only ask the robot to pick up letters that are at that moment at the robot's location and that the robot does not carry two letters in envelopes with the same color.

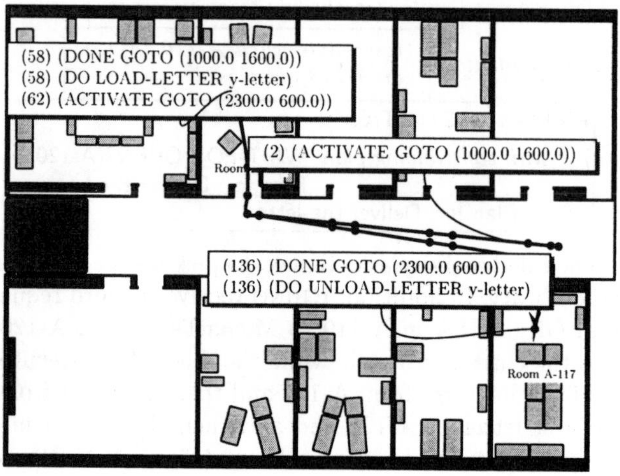


**Fig. 6.5.** A possible projected execution scenario for the initial plan. The opportunity of loading the letter of the unknown color is ignored

Suppose our robot standing in front of room A-117 has received some evidence that A-113 has been opened in the meantime. This requires the robot to reevaluate its options for accomplishing its jobs with respect to its changed belief state. Executing its current plan without modifications might yield mixing up letters because the robot might carry two letters in envelopes with the same color. The different options are: (1) to skip the opportunity, (2) to immediately ask for the letter from A-113 to be put into an envelope that is not yellow (to exclude mixing ups when taking the opportunity later); (3) to constrain later parts of the schedule such that no two yellow letters will be carried even when the letter in A-113 turns out to be yellow; and (4) to keep picking up the letter in A-113 as an opportunistic subplan. Which option the robot should take depends on its belief state with respect to the states of doors and locations of letters. To find out which schedules will probably work, in particular, which one might yield mixing up letters, the robot must apply a model of the world dynamics to the state variables.

Recall the example from the second section in which the robot standing at its initial position has perceived evidence that door A-113 has been opened. Therefore its belief state assigns probability $p$ for the value **true** of random variable open-A113. The belief state also contains probabilities for the colors of letters on the desk in A-113.

With respect to this belief state, different scenarios are possible. The first one, in which A-113 is closed, is pictured in Figure 6.5. Points on the trajectories represent predicted events. The events without labels are actions in which the robot changes its heading (on an approximated trajectory) or events representing sensor updates generated by passive sensing processes. For example, a passive sensor update event is generated when the robot passes a door. In this scenario no intervention by prediction-based debugging is necessary and no flaw is projected.

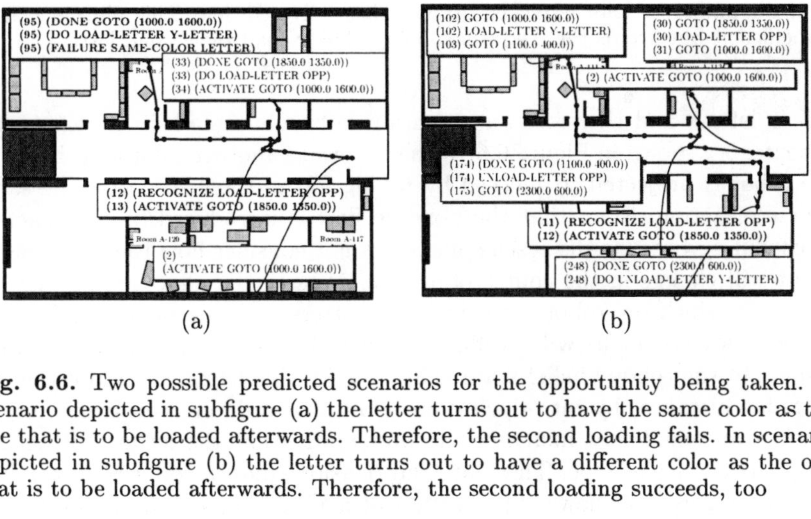


(a) (b)

**Fig. 6.6.** Two possible predicted scenarios for the opportunity being taken. In scenario depicted in subfigure (a) the letter turns out to have the same color as the one that is to be loaded afterwards. Therefore, the second loading fails. In scenario depicted in subfigure (b) the letter turns out to have a different color as the one that is to be loaded afterwards. Therefore, the second loading succeeds, too

In the scenarios in which office A-113 is open the controller is projected to recognize the opportunity and to reschedule its enabled plan steps as described above.[1] The resulting schedule asks the robot to first enter A-113, and pickup the letter for cmd-2, then enter A-111 and pick up the letter for cmd-1, then deliver the letter for cmd-2 in A-120, and the last one in A-117. This category of scenarios can be further divided into two categories. In the first subcategory shown in Figure 6.6(a) the letter to be picked up is yellow. Performing the pickup thus would result in the robot carrying two

[1] Another category of scenarios is characterized by A-113 becoming open after the robot has left A-111. This may also result in an execution failure if the letter loaded in A-113 is yellow, but is not discussed here any further.

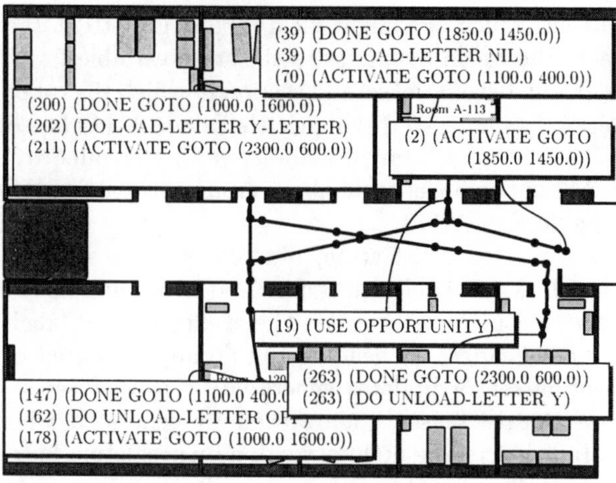


**Fig. 6.7.** Projected scenario for a plan suggested by the plan debugger. The leeter with the unknown color is picked up and also delivered first. This plan is a little less efficient but avoids the risk of not being able to load the second letter

yellow letters and therefore an execution failure is signalled. In the second subcategory shown in Figure 6.6(b) the letter has another color and therefore the robot is projected to succeed by taking for all these scenarios the same course of action. Note, that the possible flaw is introduced by the reactive rescheduling because the rescheduler does not consider how the state of the robot will change in the course of action, in particular that a state may be caused in which the robot is to carry two letters with the same color.

In this case, PPSD will probably detect the flaw if it is probable with respect to the robot's belief state. This enables the debugger to forestall the flaw, for instance, by introducing an additional ordering constraint, or by sending an email that increases the probability that the letter will be put in a particular envelope. These are the revision rules introduced in the last section. Figure 6.7 shows a projection of a plan that has been revised by adding the ordering constraint that the letter for A-120 is delivered before entering A-111.

Figure 6.8(a) shows the event trace generated by the initial plan and ***executed*** with the RHINO control system (Thrun et al., 1998) for the critical scenario without prediction based schedule debugging; Figure 6.8(b) the one with the debugger adding the additional ordering constraint. This scenario shows that reasoning about the future execution of plans in PPSD is capable of improving the robot's behavior.

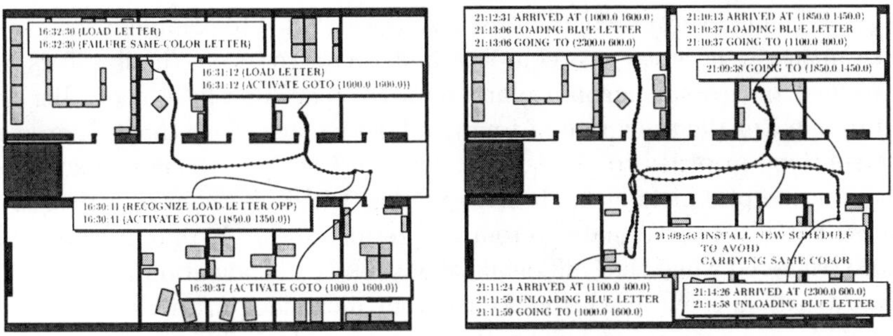


(a) (b)

**Fig. 6.8.** The trajectory without PPSD (subfigure (a)) fails because the courier did not foresee the possible complications with loading the second letter. Subfigure (b) shows a trajectory where the possible flaw is forestalled by PPSD

## 6.2 A Robot Museums Tourguide

For our second demonstration of the viability of plan-based robotic agents we have implemented an SRC as a high-level controller for an interactive museum tourguide robot called MINERVA (see Figure 6.9).

**Fig. 6.9.** The mobile robot MINERVA acting as a museum tourguide in the Smithsonian's National Museum of American History

The primary task of MINERVA's high-level controller was to schedule and execute tours such that tours took about six minutes. The museum staff estimated that the average visitor would enjoy following the robot about six minutes. Unfortunately, the time need to navigate from one exhibit to anther one depends critically on the number and the behavior of the surrounding people. The high variance of the time resources needed to complete navigation tasks makes it necessary to compose and revise tours on-the-fly. This was performed by special purpose plan adaptors. The high-level controller also had to monitor the execution of tours and detect situations that require tour interruptions in order to change the course of actions when an exceptions occurs. Examples include the battery voltage which, if below a critical level, forces the robot to terminate its tour and return to the charger. An exception is also triggered when the confidence of MINERVA's localization routines drop below a critical level, in which case the tour must temporarily be suspended to invoke a strategy for re-localization (Beetz et al., 1998).

### 6.2.1 The Plans of the Tourguide Robot

The main high-level plan of the museums tourguide robot is the one that is responsible for presenting a particular exhibit *exh*.

```
RPL procedure ACHIEVE(PRESENTED(EXH))
with local vars EXH-LOC  ←  IN-FRONT-OF(EXH)
  do
    if  EXECUTION-STATE(EXH,TO-BE-PRESENTED)
      then with valve   WHEELS   sequentially
             do in parallel do
                    ANNOUNCE-EXHIBIT(EXH)
                    ACHIEVE(LOC(RHINO, EXH-LOC))
                FACE-EXHIBIT(EXH)
                GIVE-PRESENTATION(EXH)
```

An exhibit is presented by first announcing the exhibit and navigating to a place in front of the exhibit. After having arrived at the exhibit MINERVA first faces the exhibit and then gives an oral presentation of the exhibit. The oral presentation is stored in a sound file that can be played with the sound player.

Scheduling the tourguide tours is easier than scheduling the delivery tours of the robot office courier. The tour scheduler of the museums tourguide does not have to deal with overloading failures and closed doors.

### 6.2.2 Learning Tours and Tour Management

MINERVA's plan-based controller learns a model of the time required for moving between pairs of exhibits. Learning takes place whenever the robot has successfully moved from one exhibit to another, in which case the time for the completion of the navigation task is stored and the estimate is updated using the maximum likelihood estimator. After the presentation of an exhibit is completed, MINERVA selects the exhibits and their sequence so that the whole tour best fits the desired time constraint. If the remaining time is below a threshold, the tour is terminated and MINERVA returns to the center area of the museum.

| | average | min | max |
|---|---|---|---|
| static | $398 \pm 204$ sec | 121 sec | 925 sec |
| with learning | $384 \pm 38$ sec | 321 sec | 462 sec |

**Fig. 6.10.** This table summarizes the time spent on individual tours. In the first row, tours were precomposed by static sequences of exhibits; in the second row, tours were composed on-the-fly, based on a learned model of travel time, successfully reducing the variance by a factor of 5

Table 6.10 illustrates the effect of dynamic tour revision on the duration of tours. MINERVAs environment contained 23 designated exhibits, and there were 77 sensible pairwise combinations between them (certain combinations were invalid since they did not fit together topic-wise). In the first days of the exhibition, all tours were static. The first row in Table 6.10 illustrates that the timing of those tours varies significantly (by an average of 204 seconds). The average travel time was estimated using 1,016 examples, collected during the first days of the project. The second row in Table 6.10 shows the results when tours were composed dynamically. Here the variance of the duration of a tour is only 38 seconds.

### 6.2.3 Demonstrations of the Tourguide Robot

MINERVA (see Figure 6.9) has operated for a period of thirteen days in the Smithsonian's National Museum of American History (Thrun et al., 1999). In this period, it has been in service for more than ninetyfour hours, completed 620 tours, showed 2668 exhibits, and travelled over a distance of more than fortyfour kilometers. RPL controlled MINERVA's course of action in a feedback loop that was carried out more than three times a second. Collision avoidance and self localization ran independently in faster feedback loops. MINERVA used plan adaptors for the installment of new requests, the deletion of completed plans, and tour scheduling. MINERVA reliably performed about 3200 plan adaptations.

The MINERVA experiment demonstrates that SRCs can (1) reliably control an autonomous robot over extended periods of time and (2) reliably revise plans during their execution.

## 6.3 A Robot Party Butler

The third demonstration of a plan-based robotic agent has been JAMES, an autonomous robot party butler that provided service during the reception at the Third International Conference on Autonomous Agents (AA-99) and the 23rd German Conference on Artificial Intelligence (KI-99). At the receptions, JAMES found people reception based on motion detection with a laser range finder, approached them, interacted with them, and possibly guided them to the the bar or brought them cookies. Unlike RHINO and MINERVA, JAMES is a small lowcost robot and completely controlled by its onboard computer.

## 6.4 Demonstrations of Integrated Mechanisms

The following three demonstrations describe working examples that demonstrate the intergration of conversational actions, image processing, and active localization into the high-level control system.

### 6.4.1 Communication

The following example demonstrates RHINO's communication behavior and how it is used to achieve competence in the robot's problem-solving behavior.

**Wolfram:** Could you please deliver the yellow book on the desk in room A-120 to the desk in the library before 12.30.
RHINO: Do you mean 'deliver the yellow book on Michael's desk in room A-120 to the desk in the library before 12.30' or 'deliver the yellow book on Dieter's desk in room A-120 to the desk in the library before 12.30' ?
**Wolfram:** Deliver the yellow book on Michael's desk in room A-120 to the desk in the library before 12.30.
RHINO: I will deliver the yellow book on Michael's desk in room A-120 to the desk in the library before 12:30.
**Librarian:** I will be out of the library for 45 minutes.
RHINO: OK. You will be back at 12:45.
RHINO: (sending an electronic mail to Wolfram) I can't accomplish the job "Deliver the yellow book on Michael's desk in room A-120 to the desk in the library before 12.30." The library is closed and the librarian won't be back until 12:50.

In more detail, this communication behavior is exhibited through the following problem solving steps. The first two steps show the electronic mail containing a request that is transformed into RHINO's internal representation. This representations makes the components of the speech act explicit and represents the content of the electronic mail in RHINO's command language.

**From:** peters2@cs.uni-bonn.de
**Date:** Fri, 24 Oct 1997 12:03:57
**To:** rhino+tcx@cs.uni-bonn.de
**Subject:** Library book

Could you please bring the yellow book on the desk in room a-120 to the library before 12:30

```
(REQUEST
  :SENDER (THE PERSON
                (FIRST-NAME Hanno)
                (LAST-NAME Peters))
  :RECEIVER (THE ROBOT
                (NAME RHINO))
  :TIME (THE TIME-INSTANT
                (DAY 24) (MONTH OCT)
                (TIME 12:03))
  :REPLY-WITH "Re: Library Book"
  :CONTENT content of the email
  :DEADLINE (A TIME-INSTANT
                (BEFORE
                   (TIME 12.30))))
```

The conversation plans that the RHINO control system uses implement that the request must be unambiguous with respect to the static aspects of the robot's world model and achievable. Thus if the interpretation process signals an ambiguity a control thread is triggered that leads the conversations necessary to make the request unambiguous.

After detecting the ambiguity in the request the RHINO controller generates a new speech act of the type query that contains as its content a disjunction of all instantiations of the original request that are consistent with the robot's model of the world (below right). This internal speech act representation is then transformed by the language generation process into pseudo English and the electronic mail is then sent to the person who issued the request (below right).

```
(QUERY
  :SENDER (THE ROBOT
              (NAME RHINO))
  :RECEIVER (THE PERSON
              (FIRST-NAME Hanno)
              (LAST-NAME Peters))
  :TIME (THE TIME-INSTANT ... )
  :CONTENT
    (OR (ACHIEVE
          (LOC (THE BOOK
                 (COLOR YELLOW)
                 (ON (THE DESK
                       (IN A-120)
                       (OWNER DIETER)]
               (THE LOC (IN LIBRARY)))
        (ACHIEVE
          (LOC (THE BOOK
                 (COLOR YELLOW)
                 (ON (THE DESK
                       (IN A-120)
                       (OWNER MICHAEL)]
               (THE LOC (IN LIBRARY)))
  :DEADLINE
    (A TIME-INSTANT (BEFORE 12.30)))
```

**From:** Rhino (rhino+tcx)
**Date:** Fri, 24 Oct 1997 12:04:35
**To:** peters2@cs.uni-bonn.de
**Subject:** Your Command

**Do You mean**
**achieve that** the yellow book on dieter's desk in room A-120 is in the library before 12.30 **will be true**
**or**
**achieve that** the yellow book on michael's desk in room a-120 is in the library before 12.30 **will be true**

Finally, the person issuing the request picks the intended instantiation of the original request and sends it back to RHINO. RHINO acknowledges the receipt of the request by repeating its interpretation of the request to be performed.

**From:** Hanno Peters
**Date:** Fri, 24 Oct 1997 12:05:48
**To:** rhino+tcx@cs.uni-bonn.de
**Subject:** Command-Revision

**achieve that** the yellow book on michael 's desk in-room a-120 is in the library before 12.30 **will be true.**

**From:** Rhino (rhino+tcx@cs.uni-bonn.de)
**Date:** Fri, 24 Oct 1997 12:08:35
**To:** peters2@cs.uni-bonn.de
**Subject:** Your Command

**I will achieve that** the yellow book on michael 's desk in-room a-120 is in the library before 12.30 **will be true.**

### 6.4.2 Image Processing

Let us consider another demonstration showing the advantages accruing from having image processing as an integral, explicit and transparent part of the control system. We consider the classic task of visual place recognition and the recognition of known pieces of furniture within an office environment. In this problem context, specialization and adaptation to the environment promise great simplifications of the visual processing required. The image processing functionality for this type of task is well known, permitting us to focus on the added functionality which comes from employing SRIPPs. Simply put, RHINO employs a geometric projection algorithm that takes RHINO's position and the direction in which the camera is pointing as arguments, retrieves the furniture items in the robot's view, and predicts where the corners of the furniture item should appear in the camera image. Thus RHINO applies a compound image procesing routine which searches for the predicted lines suggested by the geometric projection algorithm.

RHINO is responsible for the control of the image procesing operations required for object recognition. Since these operations are represented as plans, RHINO can tailor the visual recognition of furniture items by selecting the lines to be found and by sorting these lines in an order that should promise fast and reliable recognition. As an example of specialization, suppose that the color of the item of furniture differs significantly from the background. The object recognition procedure can then look for the boundary corners first. If, for example, proximity sensors detect an object standing in front of the item of furniture, the process can adapt its plan, and thus its sequence of image processing operations, to prefer vertical lines and high corners. The procedure will run the routines for detecting and recognizing lines in parallel or sequentially depending on CPU load. The confidence value for a successful recognition will also be set according to specified deadlines for the recognition process. In the case of recognition failures the robot control program running the recognition procedure can try to relocalize the robot or move the robot to a different location near the furniture item to perhaps get a better view.

Figure 6.11 shows the output of different steps of the recognition procedure. In Figure 6.11(a) one can see the region of interest (black) that is

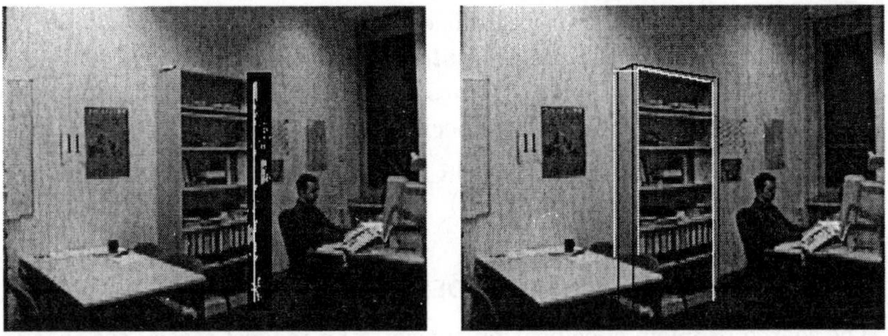

(a) (b)

**Fig. 6.11.** Recognizing the shelf using runtime configured image processing plans. The robot uses context information including its estimated pose, the accuracy of the pose estimation, and the shelf position, to predict where the long vertical lines of the shelf should appear in the image. It generates regions of interest to focus the image processing routines

computed from the degree of confidence in the robot's location with the white lines being the edges detected by the edge detector within this region. Figure 6.11(b) shows the final result of recognizing the shelf. The black lines are the predicted ones and the white ones are the recognized lines. Such results are to be expected from this type of IP.

It would be impossible to achieve this level of flexibility and robustness if image processing were hidden within discrete and fixed image processing modules that could be used only in a very limited and predefined way and were controlled by stored sequences of image processing requests. Since image processing routines in our approach are dynamically generated and revised during robot operation, the robot control system needs direct access to the visual information sensed by the robot. Moreover, since the control routines are concurrent programs, they may simultaneously need to read or modify the shared visual information. In this context, the image processing facility must enforce controlled access to the visual information to ensure stability and optimal use of available resources.

### 6.4.3 Resource-Adaptive Search

Despite the fact that search tasks are ubiquitus in service robot applications surprisingly little work has been done on developing computational models capable of producing competent search behavior. We will now describe a search method and its integration into SRCs, which exploits background knowledge and considers given deadlines to tailor the robot's search behavior to the particular search task at hand. The system generates search plans for a small-size office environment based on cost-benefit considerations within seconds.

A typical search task of a robot office courier can be paraphrased as: *"look for the red letter, I need it in ten minutes. I probably left it in room A-120 or possibly in A-111. I usually leave things around my desk and sometimes next to Dieter's."* Such a search task is specified as follows.

**REQUEST** FIND (A LETTER (COLOR RED))
:DEADLINE (+ NOW 10))
**TELL** P(IN-ROOM = A-120) = 0.4
∧ P(IN-ROOM = A-111) = 0.4
∧ P(NEXT-TO = MICHAELS-DESK | IN-ROOM = A-120) = 0.6) ∧ ...

The search problem specification consists of a request statement that tells the robot to find a red letter within ten minutes and the tell statement that provides information about where the robot can expect to find a red letter. It specifies that the probability of finding a red letter in room A-120 is 0.4 (the random variable IN-ROOM has the value A-120 with the probability of 0.4) and that the probability of a red letter lying next to Michael's desk given that the letter is in room A-120 is 0.6.

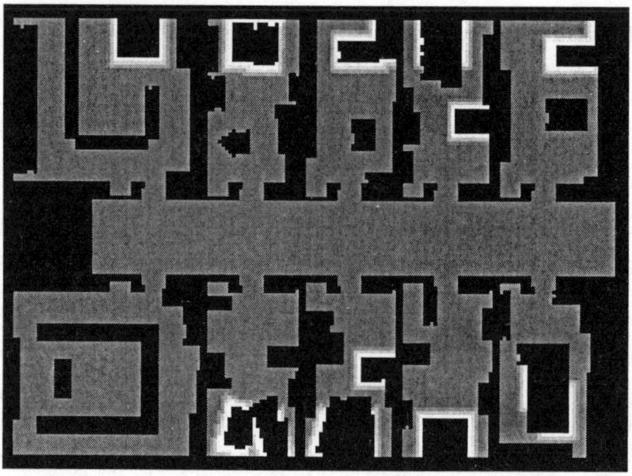

**Fig. 6.12.** The figure shows a probability distribution for a target location that is constrained by a search task. The brighter a grid cell the higher the probability that the target object is at that location

Spatial relations such as "next to" specify conditions on object configurations, for example on the distance between the target, a letter, and a reference object, Michael's desk. We define the probability of an object being at location ⟨x,y⟩ if the object is in front of a map object as a bivariate Gaussian probability distribution (see (Gapp, 1994) for a more detailed account and an experimental justification). The information content of the background information provided with a search problem is translated into a probability distribution for the location of the target (see figure 6.12).

Figure 6.13 pictures four search plans that a plan adaptor of an SRC has generated for different kinds of search problems. The first one (figure 6.13(a)) searches a Coke can without time constraints and a priori information. The robot searches the whole environment systematically in order to minimize the expected time to find the object. Figure 6.13(b) pictures a search behavior without resource constraints but with prior information that the Coke can is either in the bottom leftmost or the two upper leftmost rooms. A search plan for time-constrained search is shown in figure 6.13(c). Finally, the last plan is computed for time-constrained, informed search. The Coke can is in one of the rooms listed above. The difference to the last search plan is that the robot does not search the hallway. Instead it searches the upper leftmost room more thoroughly.

**Fig. 6.13.** Problem-specific search plans generated by the resource-adaptive, informed search strategy. Subfigure (a) shows an uninformed search plan that searches the region systematically. Subfigure (b) depicts a search path where the robot is told the the target is in one of the three rooms. A search plan that has maximum probability of finding the target within 15 minutes is shown in subfigure (c). Finally, a search path that only searches in three rooms and for 15 minutes is shown in subfigure (d)

The search routines are integrated into the SRC of the robot office courier. The search routine is essentially a loop that carries out one search action after the other, where a search action consists of navigating to the viewpoint of the search grid cell, orienting the robot in the direction of the center of the search grid cell, and running the target detection procedure. Concurrently with the execution of the main loop are monitoring processes that detect situations such as the way to the intended viewpoint is blocked or the door to a room that should be searched is closed. If the way to a viewpoint is blocked the search action recovers locally and by trying an alternative viewpoint (see figure 6.14). If no alternative viewpoint is possible the search action fails and the search action of the next iteration is executed. If doors are detected to be closed the intended search plan is replanned under the evidence that the particular room is closed.

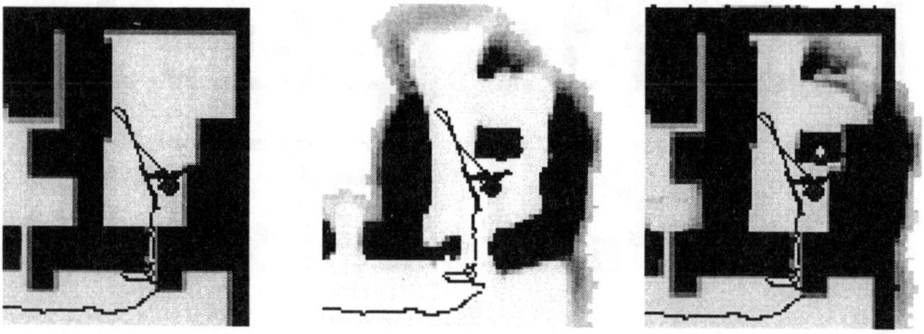

(a) (b) (c)

**Fig. 6.14.** Replanning search when viewpoints are blocked

Figure 6.15 shows how a search plan is replanned if a door to a room to be searched is closed. Figure 6.15(a) shows the execution of the plan if the library is open. The right one shows the path if the library is closed. In this case the classroom is searched more thoroughly.

### 6.4.4 Active Localization

In this experiment we show how SRCs robustly integrate the subtask of estimating the position of the robot in the robot's operation. The principle of this integration is to monitor the certainty of the position estimation and to autonomously relocalize the robot whenever the uncertainty grows too large. A localization process keeps accurately and efficiently track of the robot's position (Burgard, Fox, and Thrun, 1997). Furthermore, it provides a measure for detecting localization failures and it is able to autonomously relocalize the robot in such situations. The SRC ensures that the robot's missions can

Fig. 6.15. Replanning the search when a door is closed

be interrupted by such active localization processes at any point in time and allow the robot to complete its mission afterwards.

To demonstrate these capabilities we performed several complex experiments in our office environment. In all these experiments the task of the robot was to reach a location in front of a shelf in one room of our building. In order to force the control system to start the active localization process we manually rotated the robot by about 70 degrees shortly after starting its mission.

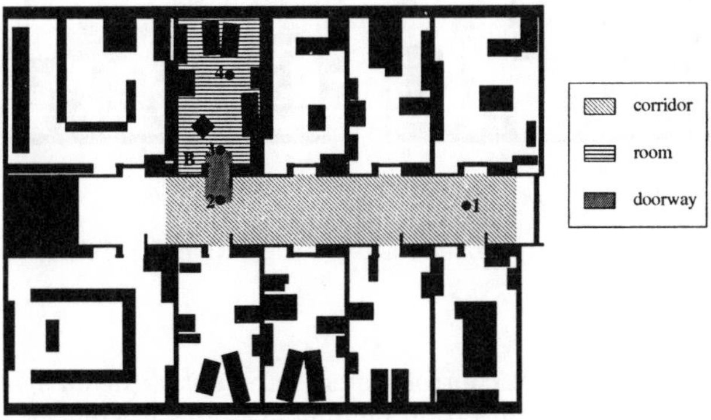


Fig. 6.16. Navigation plan of the robotic agent

Figures 6.16-6.18 show a representative example of these experiments. Figures 6.17-6.18 contain an outline of our environment including the corresponding position probability density (dark regions are more likely). In this example RHINO was started at the east end of the hallway (see Figure 6.17(a)). The destination was a location in front of Wolfram's shelf in room B. Upon receiving the request to go in front of Wolfram's shelf, the structured reactive controller computed the navigation plan shown in Figure 6.16. Shortly after starting the execution of the navigation plan we manually turned the robot to the right. After this turn the robot was facing west (i.e. to the left of the page) while the position tracking assumed the robot to face south-west-south. Consequently the path planner generated target points that caused the robot to turn right and move ahead. Shortly after that, RHINO's reactive collision avoidance (Fox, Burgard, and Thrun, 1997) turned the robot to the left because of the northern (top) wall of the corridor. While the solid line in Figure 6.17(a) shows the real trajectory of the robot, the dashed line depicts the trajectory estimated by the position estimation module. Note that this module didn't realize the manual change in orientation. To the robot it seemed as if there was some obstruction in the corridor which forced it off course and through the doorway. At the end of this path the accumulated discrepancies between expected and measured sensor readings reduced the confidence in the position of the robot below the given threshold.

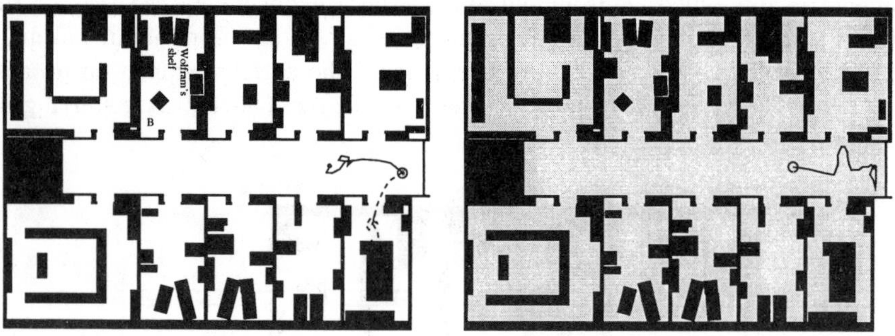


(a) (b)

**Fig. 6.17.** Path of the robot before active localization (a). Initial stage of active localization (b)

At that point the robot control system interrupted the navigation process and started the active localization module. The inconsistencies in the belief state caused the localization module to re-localize the robot, which starts with uniform distribution over all possible positions in the environment (see Figure 6.17(a)). After a short period of random movement, the number of possible positions had been reduced to 10 local maxima in the belief state where the positions at the west end facing east and the east end facing west were the most likely ones. To disambiguate these two positions, RHINO decided to enter the next room to its left (either room C or D). This target location was chosen because it was expected to maximally reduce the uncertainty of the position estimation. In this particular run the corresponding navigation action lead RHINO into room C, where in fact all ambiguities could be resolved (see Figure 6.18(a)).

After successful relocalization, the control was transferred back to the navigation task which generated a new navigation plan to room B, which as shown in Figure 6.18(b) was executed immediately after re-localization. Figure 6.18(b) also shows the complete trajectory taken by the robot.

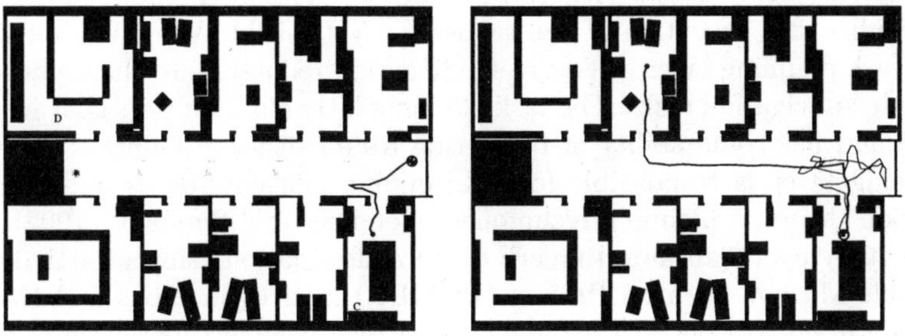


**Fig. 6.18.** RHINO moves into room C for disambiguation (a). Navigation after successful re-localization (b)

This demonstration shows how monitoring processes, recovery activities, and navigation plans are integrated. It demonstrates how navigation tasks can be interrupted and successfully continued despite the side effects of the interrupting subtasks.

## 6.5 Related Work on Plan-Based Robotic Agents

In recent years, there have been a number of longterm real-world demonstrations of plan-based control, which have impressively shown the potential impact of this technology for future applications of autonomous service robots. In NASA's Deep Space program a plan-based robot controller, called the *Remote Agent* (Muscettola et al., 1998b), has autonomously controlled scientific experiments in the space. In the *Martha* project (Alami et al., 1998), fleets of robots have been effectively controlled and coordinated. *Xavier* (Simmons et al., 1997a), an autonomous mobile robot with a plan-based controller, has navigated through an office environment for more than a year, allowing people to issue navigation commands and monitor their execution via the Internet. In 1998, *Minerva* (Thrun et al., 1999), another plan-based robot controller, acted for thirteen days as a museum tourguide in the Smithonian Museum, and led several thousand people through the exhibition.

In this section, we will describe and discuss several of these autonomous robots in more detail.

### 6.5.1 XAVIER

Simmons et al. (1997a) have developed XAVIER, an autonomous mobile robot that has performed longterm navigation experiments in an office building. XAVIER is implemented on the basis of a software architecture (Simmons, 1994) that is organized in five layers: obstacle avoidance, navigation, path planning, task scheduling, and the user interface (Simmons et al., 1997b). The user interface allows users to post asynchronous navigation requests and to monitor the execution of the requests via a World Wide Web interface. The task planning layer inserts new navigation requests into the sequence of intended navigation tasks. The task planning layer also monitors goal achievement and has some ability of recovering from execution failures. The path planning layer is responsible for determining efficient routes and employs decision-theoretic planning techniques (Goodwin and Simmons, 1992). The navigation layer plans how to reach the next destination using a Partially Observable Markov Decision Process (POMDP) model (Simmons and Koenig, 1995). Finally, the obstacle avoidance layer reactively directs the robot to the next destination guided by the navigation plan.

XAVIER has been in nearly daily use for more than one year and traveled in this period more than 75 kilometers in order to satisfy over 1800 navigation requests.

The layer that is closest related to the research presented this thesis is task planning layer. At the task planning layer, XAVIER employs ROGUE, a system for interleaved task planning, execution, and situation-dependent learning (Haigh and Veloso, 1997). ROGUE accepts multiple, asynchronous task requests, and interleaves task planning with real-world robot execution. ROGUE prioritizes tasks, suspends and interrupts tasks, and opportunistically achieves compatible tasks. ROGUE also monitors task execution and compensates for failure and changes in the environment.

Our approach to plan-based control differs from the approach taken in the XAVIER controller in various respects. First, XAVIER is only concerned with navigation tasks, whereas our system is concerned with the concurrent and integrated control of a variety of mechanisms. Whereas in XAVIER task management is performed by a module of the control system, SRCs perform plan management by explicitly specified plan adaptors, constraining plans that are part of the robot's overall plan. Managing plans through plans rather than a blackbox system has several advantages. First, plan management can be controlled in the same way as perception actions. Second, the plan adaptors being explicit and rather declarative is a prerequisite for successfully learning plan adaptors, their applicability, and expected utility. Third, we can realize situation specific plan management strategies by having plan adaptors that insert, delete, and revise other plan adaptors.

ROGUE is very interesting because of its means for learning the situation-specific cost of performing actions (Haigh and Veloso, 1998). Learning cost models of the actions of a robotic agent is certainly an important step of efficiently performing everyday activity. We plan to investigate the learning of more reliable navigation plans and the situation-specific cost of navigation actions.

### 6.5.2 CHIP

Firby et al. (1996) have developed CHIP, an autonomous mobile robot that is intended to serve as a general-purpose robotic assistant. A main application demonstration of CHIP was the Office Cleanup event of the 1995 Robot Competition and Exhibition. CHIP was to scan an entire area systematically and, as collectible objects were identified, pick them up and deposit them in the nearest appropriate receptacle.

The software architecture of CHIP's control system is called the Animate Agent Architecture (Firby et al., 1995). The Animate Agent Architecture consists of two components: the RAP reactive plan executor (Firby, 1987; Firby, 1989), and a control system based on reactive skills (Firby, 1992; Firby, 1994). The executor executes sketchy plans, that is, plans that leave many steps to be chosen as the actual situation unfolds. It makes these choices at run time when the true state of the world can be sensed. The output from the executor is a detailed set of instructions for conguring the skills that make use of the agent's sensors and eectors.

One of the primary objectives in the CHIP project is to develop a set of skills, which closely correspond to our notion of process modules, and subgoals that are generic and can be used to achieve a wide variety of everyday tasks. Their work is very interesting because they have developed a library of low-level plans that can be carried out in situation-specific ways. In developing control plans for real robot applications they have encountered many issues that we have also addressed in our chapter on plan representation.

Their work is closely related to our work on integrating mechanisms and low-level plan representation. Indeed, their notion of robot skills closely resembles the notion of process modules that we have employed in this Thesis. Their notion of reactive action packages roughly corresponds to our low-level plans.

There are several distinguishing aspects between both research projects. The ANIMATE AGENT PROJECT has investigated robotic tasks that included object manipulation and more sohisticated vision tasks (Prokopowicz et al., 1996). Manipulation tasks raise a number of interesting issues for plan-based control, including the tight integration of image processing and manipulation, in particular grasping, the necessity to parameterize manipulation actions with object descriptions, and the occurrence of perceptual confusion. Object manipulation seems to present many realistic problems that call for more sophisticated plan management mechanisms. While the scope of mechanisms that the plans control is wider, the set of plan management operations is more or less restricted to the situation-specific plan expansion.

### 6.5.3 Flakey

Konolige et al. (1997) have developed FLAKEY, a custom-built robot platform that operates within SRI's office environment. FLAKEY navigates through an office environment according to a multistep strategic plan, an approximate map of the area, and the information provided by several on-board sensors. The robot executes the plan despite the presence of unanticipated objects, moving obstacles (e.g., people), and noisy sensors.

FLAKEY's software architecture, called SAPHIRA (Konolige et al., 1997), is a layered control system consisting of a effector level, a behavior level, and a task level. In SAPHIRA sensor data and their interpretations in terms of perceptual features are stored in a database that represents the robot's egocentric view. The database is called the local perceptual space (LPS). On the basis of an approximate map of the world, the robot matches (anchoring) its observations to objects in its environment.

At the behavior level, FLAKEY is controlled by behaviors, which in turn are realized through a set of fuzzy rules whose antecedents are fuzzy formulas (Saffiotti et al., 1993). A rule's level of activation corresponds to the degree that its precondition matches the LPS; when the precondition of a rule holds, then its action is very desirable; when the precondition partially matches, the action is less desirable. By blending the actions of all the rules, in the currently active behaviors, according to their levels of activation, an "optimal" action is derived. Since the conditions in the environment typically change slowly, the repeated application of the same fuzzy rules smoothly blends actions over time.

At the task level, PRS-Lite (Myers, 1996) provides real-time supervisory control for the robot by managing the robot's intentions according to the beliefs and goals of the robot. A large part of the beliefs are retrieved from the LPS. PRS-Lite follows a strategic plan that states the sequence of goals that the robot needs to satisfy. PRS-Lite utilizes a library of predefined procedures to reduce these high-level goals into subsequences of lower-level goals; at the lowest level, goals are satisfied by activating and deactivating behaviors according to the refined plan and the situation-dependent information in the LPS. Thus, at any point in tim, several behaviors are enabled, with their level of activation determined by the fuzzy controller.

The unique feature of PRS-Lite is that it produces smooth transitions between different sets of behaviors. Smooth transitions are achieved by terminating tasks after the subsequent tasks are enabled and letting the fuzzy controller lower the activation level of the previous and raise the activation level of the following behavior set.

Our approach to plan-based control of robotic agents differs from the FLAKEY approach in several ways. FLAKEY dispenses with execution time deliberation and can therefore not predict and forestall execution failures in the way Structured Reactive Controllers do. In another research project, however, PRS-Lite has been coupled with an AI planning system (Wilkins et al., 1995). Another difference is the way task-level control is specified. While in SRCs the intended course of action is specified by a concurrent reactive plan, PRS-Lite dynamically expands the intended course of action according to the actual beliefs.

### 6.5.4 An Architecture for Autonomy

Alami et al. (1998) have developed an integrated software architecture for autonomous robot control that allows a mobile robot to plan its tasks, taking into account complex temporal, resource, and other domain constraints. The architecture is composed of two main levels: a decision level, which is both goal- and event-driven, and a functional execution level composed of modules that perform state estimation and effector control. The decision level itself is typically composed of a mission planning and a task supervisor and integrates deliberation and reaction.

The basic components of the software architecture are IxTeT (Laborie and Ghallab, 1995; Ghallab and Laruelle, 1994), a temporal planner, PRS (Ingrand et al., 1996; Ingrand, Georgeff, and Rao, 1992), for task refinement and supervision, Kheops for the reactive control of the functional level, and GenoM for the specification and integration of modules at the functional level (Fleury, Herrb, and Chatila, 1994). Instances of the proposed architecture have been already integrated into several indoor and outdoor robots.

A salient aspect of the architecture is the use of a sophisticated temporal planner, called IxTeT (Laborie and Ghallab, 1995; Ghallab and Laruelle, 1994), that can synthesize complex mission plans that satisfy the symbolic and numerical constraints for time points and constraints on resources and other domain attributes. The research on IxTeT focuses on the investigation of effective compromises between the expressiveness of action and subplan description languages and the efficiency of the search for solutions of the planning problems. In this context, the IxTeT makes important contributions to control issues involved in synthesizing complex plans. The only other robotic agent architecture that is equipped with such a sophisticated mission planning mechanism is the Remote Agent that is discussed in the next section.

Another important aspect of the Architecture for Autonomy that distinguishes it from other architectures is its emphasis on software engineering tools for system design and implementation, most notably GenoM (Fleury, Herrb, and Chatila, 1994), Proprice (Despouys and Ingrand, 1999), and TransGen. Alami et al. (2000) describe an experiment in which they have integrated existing low-level modules into a complete control system that performs nontrivial task control within 40 days.

A key application of the architecture is the Martha project (Alami et al., 1998), which investigates the management of a fleet of autonomous mobile robots for transshipment tasks in harbours, airports and marshaling yards. In such context, the dynamics of the environment, the impossibility to correctly estimate the duration of actions (the robots may be slowed down due to obstacle avoidance or re-localisation actions, and delays in load and unload operations, and so on) prevent a central system from elaborating long or medium term efficient and reliable detailed robot plans.

In the Martha project, a more flexible way is pursued in which the robots determine incrementally the resources they need taking into account the execution context. As a consequence, Alami, Ingrand, and Qutub (1998) have developed advanced techniques for plan merging in order to improve multi robot cooperation.

### 6.5.5 Remote Agent

The REMOTE AGENT is a software architecture for controlling autonomous spacecrafts (Muscettola et al., 1998b). The REMOTE AGENT has successfully operated the Deep Space 1 spacecraft for two days in May of 1999. This experiment was the first time that an autonomous agent has controlled a deep space mission.

The control of spacecrafts presents autonomous robot controllers with a unique combination of challenges. These challenges comprise the requirement for autonomous operation over long periods of time, high reliability including the capability of fast failure response despite limited observability of the spacecraft state.

The REMOTE AGENT (Pell et al., 1997) consists of three components: a smart executive (Pell et al., 1997a; Pell et al., 1998), a mode-identification and reconfiguration engine (Williams and Nayak, 1996), and an integrated planner and scheduler (Pell et al., 1997b). The REMOTE AGENT is commanded by giving it a small handful of high-level goals. The planner/scheduler expands these goals into a plan that achieves the goals while satisfying constraints on resources, operations, and safety. The plan is carried out by the executive. If problems arise during execution, they are detected by mode identification module and the executive tries to resolve the problems within the constraints of the plan. If the problems cannot be resolved without violating the constraints, the executive terminates the current plan, achieves a safe spacecraft state, and asks for a new plan. The planner generates a new plan that achieves the remaining goals while taking the new spacecraft state into account.

Key design principles underlying the REMOTE AGENT are the use of *model-based programming*, the application of *deduction and search* mechanisms even in fast feedback loops (Williams and Nayak, 1997), and the use of *high-level closed loop commanding*. The basic idea of model-based programming is the implementation of control programs based on declarative component models of hardware and software behaviors. Using the models, the REMOTE AGENT can reason about system wide interactions between components. Even the procedures of the reactive execution component operate with these declarative models. The model-based execution system supports diverse reasoning tasks by the use of self models of modules and mechanisms for self configuration. Such self models enable components to identify anomalous and unexpected situations and to synthesize correct responses to requests. The advantages of model-based autonomy include facilities for more efficient programming of system functionality and increased safety and reliability.

The need for onboard and execution time deduction and search arises from the controlled system being extremely complex and yielding system wide interactions. Because of these complex interactions it is in general not possible to precompute the appropriate responses to all possible combinations of system states.

Finally, high-level closed-loop commanding is employed in order to avoid the premature committment to overspecialized plans. Thus, the REMOTE AGENT specifies goal trajectories that specify goals and time windows in which they should be achieved. Such goal trajectories give the agent more flexibility with respect to the specific realizations of the goals.

## 6.6 Discussion

We have seen in this chapter examples in which modern AI planning technology in general and Structured Reactive Controllers in particular have contributed to autonomous robot control by improving the robot's behavior. The planning techniques, in particular, the temporal projection and the plan revision techniques can do so because they (1) extend standard scheduling techniques and reactive plan execution techniques with means for predicting that the execution of a scheduled activity will result in a behavior flaw; (2) predict states relevant for making scheduling decisions that the robot won't be able to observe; (3) uses information about predicted states before the robot can observe them.

These capabilities enable planning robots to outperform those that have controllers that do only situated task scheduling. However, in simple environments as the one we have considered in this thesis the number of scenarios in which in which planning pays off is surprisingly small. The prediction of flaws could not be exploited convincingly because the only possible flaw that could be detected this way was the loading of two letters with the same color. For such simple problem and dynamics one could use a special purpose detector. The second capability could be exploited because the environment was too small and the range of sensors too large. The third capability to predict whether doors will probably be open before they can be observed could not be exploited because it is possible to wait until the robot observes the door without losing performance.

Our conclusion from these lessons is that the break-even point for the application of robot action planning techniques will be reached when service robots will (1) perform their jobs in environments that are more complex, (2) will be able to perceive more states that can be used for planning, and (3) have to deal with more things that might go wrong as they perform their activities.

# 7. Conclusions

Our long-term research goal is to understand and build autonomous robot controllers that can carry out daily jobs in offices and factories with a reliability and efficiency comparable to people. We believe that behavior patterns, such as exploiting opportunities, making appropriate assumptions, and acting reliably while making assumptions, make everyday activity efficient and reliable. Exhibiting such behavior patterns requires robot controllers to specify concurrent, reactive behavior — even at the deliberative level of agent control.

In this book we have proposed SRCs as a means for specifying the behavior of autonomous service robots. SRCs can accomplish important patterns of everyday activity successfully because their sub-plans are made interruptable and restartable using high-level control structures that specify synchronized concurrent reactive behavior. SRCs achieve adaptivity through plan revision and scheduling processes, implemented as policies, that detect opportunities, contingent situations, and invalid assumptions. Plan revision techniques can perform the required adaptations because of the modular and transparent specification of concurrent and reactive behavior. In particular the distinction of plan adaptors and primary activities increases the modularity significantly. Plan adaptors enable RHINO to specify opportunistic behavior and achieve reliable operation while making simplifying assumptions.

One conclusion that we draw is that there is no fundamental trade-off between the expressiveness of the language for specifying robot controllers and the feasibility of automatically reasoning about and revising these controllers. We can use expressive languages and still design the plans in ways that allow for fast reasoning and reliable revision. This suggests the exploration of new variations of planning problems that might be better suited for autonomous robots control. Our most important research direction aims at getting a more thorough understanding about the design principles, why and when they work, and a better understanding on the impact of these principles on the complexity of reasoning problems. Other ongoing research projects on structured reactive controllers investigate methods for learning routine plans from experience and the development of plan revision policies that forestall plan failures based on prediction.

# Bibliography


Alami, R., R. Chatila, S. Fleury, M. Ghallab, and F. Ingrand (1998). An architecture for autonomy. *International Journal of Robotics Research* 17(4).

Alami, R., R. Chatila, S. Fleury, M. Herrb F. Ingrand, M. Khatib, B. Morisset, P. Moutarlier, and T. Simeon (2000). Around the lab in 40 days ... In *Proceedings of the IEEE International Conference on Robotics and Automation (ICRA 2000).*

Alami, R., S. Fleury, M. Herb, F. Ingrand, and F. Robert (1998). Multi robot cooperation in the Martha project. *IEEE Robotics and Automation Magazine* 5(1).

Alami, R., F. Ingrand, and S. Qutub (1998). A scheme for coordinating multi-robots planning activities and plans execution. In Prade, H., editor, *Proceedings of the 13th European Conference on Artificial Intelligence (ECAI-98)*, pp. 617–621, Chichester. John Wiley & Sons.

Allen, J. and G. Ferguson (1994). Actions and events in interval temporal logic. *Journal of Logic and Computation* 4(5): 531–579.

Allen, J., H. Kautz, R. Pelavin, and J. Tenenberg, editors (1990). *Reasoning about Plans.* Morgan Kaufmann, San Mateo, Cal.

Aloimonos, J., I. Weiss, and A. Bandopadhay (1987). Active vision. *International Journal on Computer Vision* pp. 333–356.

Alur, R., T. Henzinger, and P. Ho (1996). Automatic symbolic verification of embedded systems. *IEEE Transactions on Software Engineering* 22(3): 181–201.

Alur, R., T. Henzinger, and H. Wong-Toi (1997). Symbolic analysis of hybrid systems. In *Proceedings of the 37th IEEE Conference on Decision and Control.*

Andre, D. and S. Russell (2001). Programmable reinforcement learning agents. In *Proceedings of the 13th Conference on Neural Information Processing Systems*, pp. 1019–1025, Cambridge, MA. MIT Press.

Arbuckle, T. and M. Beetz (1998). RECIPE - a system for building extensible, run-time configurable, image processing systems. In *Proceedings of Computer Vision and Mobile Robotics (CVMR) Workshop*, pp. 91–98.

Arbuckle, T. and M. Beetz (1999). Extensible, runtime-configurable image processing on robots — the RECIPE system. In *Proceedings of the 1999 IEEE/RSJ International Conference on Intelligent Robots and Systems.*

Arkin, R. (1998). *Behavior based Robotics.* MIT Press, Cambridge, Ma.

Arkin, R. C. (1991). Integrating behavioral, perceptual, and world knowledge in reactive navigation. In Maes, P., editor, *Designing Autonomous Agents*, pp. 105–122. MIT Press, Cambridge, MA, USA.

Austin, J. L. (1962). *How To Do Things with Words.* Harvard University Press, Cambridge, Massachusetts.

Bacchus, F., J.Y. Halpern, and H. Levesque (1999). Reasoning about noisy sensors and effectors in the situation calculus. *Artificial Intelligence 111(1-2)* .

Bajcsy, R. (1988). Active perception. *Proceedings of IEEE* 76: 996–1005.

Ballard, D. (1991). Animate vision. *Artificial Intelligence* 48: 57–86.

Barbuceanu, M. and M. Fox (1995). COOL: A language for describing coordination in multiagent systems. In Lesser, V., editor, *Proceedings of the First International Conference on Multi-Agent Systems*, pp. 17–24, San Francisco, CA. MIT Press.

Beetz, M. (1999). Structured Reactive Controllers — a computational model of everyday activity. In Etzioni, O., J. Müller, and J. Bradshaw, editors, *Proceedings of the Third International Conference on Autonomous Agents*, pp. 228–235.

Beetz, M. (2000). *Concurrent Reactive Plans: Anticipating and Forestalling Execution Failures*, Vol. LNAI 1772 of *Lecture Notes in Artificial Intelligence*. Springer Publishers.

Beetz, M. (2001). Structured Reactive Controllers. *Journal of Autonomous Agents and Multi-Agent Systems. Special Issue: Best Papers of the International Conference on Autonomous Agents '99* 4: 25–55.

Beetz, M., T. Arbuckle, T. Belker, M. Bennewitz, A. Cremers, D. Hähnel, and D. Schulz (2000). Enabling autonomous robots to perform complex tasks. *KI - Künstliche Intelligenz: Special Issue on Autonomous Robots* .

Beetz, M., T. Arbuckle, M. Bennewitz, W. Burgard, A. Cremers, D. Fox, H. Grosskreutz, D. Hähnel, and D. Schulz (2001). Integrated plan-based control of autonomous service robots in human environments. *IEEE Intelligent Systems* 16(5): 56–65.

Beetz, M., T. Arbuckle, A. Cremers, and M. Mann (1998). Transparent, flexible, and resource-adaptive image processing for autonomous service robots. In Prade, H., editor, *Procs. of the 13th European Conference on Artificial Intelligence (ECAI-98)*, pp. 632–636.

Beetz, M. and T. Belker (2000a). Environment and task adaptation for robotic agents. In Horn, W., editor, *Procs. of the 14th European Conference on Artificial Intelligence (ECAI-2000)*.

Beetz, M. and T. Belker (2000b). Learning structured reactive navigation plans from executing mdp navigation policies. In Ferryman, editor, *8th International Symposium on Intelligent Robotic Systems, SIRS 2000*.

Beetz, M. and M. Bennewitz (1998). Planning, scheduling, and plan execution for autonomous robot office couriers. In Bergmann, R. and A. Kott, editors, *Integrating Planning, Scheduling and Execution in Dynamic and Uncertain Environments*, Vol. Workshop Notes 98-02. AAAI Press.

Beetz, M., M. Bennewitz, and H. Grosskreutz (1999). Probabilistic, prediction-based schedule debugging for autonomous robot office couriers. In *Proceedings of the 23rd German Conference on Artificial Intelligence (KI'99), Bonn, Germany*. Springer Verlag.

Beetz, M., W. Burgard, D. Fox, and A. Cremers (1998). Integrating active localization into high-level control systems. *Robotics and Autonomous Systems* 23: 205–220.

Beetz, M., M. Giesenschlag, R. Englert, E. Gülch, and A. B. Cremers (1999). Semi-automatic acquisition of symbolically-annotated 3d models of office environments. In *International Conference on Robotics and Automation (ICRA-99)*.

Beetz, M. and H. Grosskreutz (1998). Causal models of mobile service robot behavior. In Simmons, R., M. Veloso, and S. Smith, editors, *Fourth International Conference on AI Planning Systems*, pp. 163–170, Morgan Kaufmann.

Beetz, M. and H. Grosskreutz (2000). Probabilistic hybrid action models for predicting concurrent percept-driven robot behavior. In *Proceedings of the Sixth International Conference on AI Planning Systems*, Toulouse, F. AAAI Press.

Beetz, M. and D. McDermott (1992). Declarative goals in reactive plans. In Hendler, J., editor, *First International Conference on AI Planning Systems*, pp. 3–12, Morgan Kaufmann.

Beetz, M. and D. McDermott (1994). Improving robot plans during their execution. In Hammond, K., editor, *Second International Conference on AI Planning Systems*, pp. 3–12, Morgan Kaufmann.

Beetz, M. and D. McDermott (1996). Local planning of ongoing activities. In Drabble, Brian, editor, *Third International Conference on AI Planning Systems*, pp. 19–26, Morgan Kaufmann.

Beetz, M. and D. McDermott (1997). Expressing transformations of structured reactive plans. In *Recent Advances in AI Planning. Proceedings of the 1997 European Conference on Planning*, pp. 64–76. Springer Publishers.

Beetz, M. and H. Peters (1998). Structured reactive communication plans — integrating conversational actions into high-level robot control systems. In *Proceedings of the 22nd German Conference on Artificial Intelligence (KI 98), Bremen, Germany.* Springer Verlag.

Belker, T., M. Beetz, and A. Cremers (2002). Learning action models for the improved execution of navigation plans. *Robotics and Autonomous Systems* .

Blythe, J. (1995). AI planning in dynamic, uncertain domains. In *Extending Theories of Action: Formal Theory & Practical Applications: Papers from the 1995 AAAI Spring Symposium*, pp. 28–32. AAAI Press, Menlo Park, California.

Blythe, J. (1996). Decompositions of Markov chains for reasoning about external change in planners. In Drabble, B., editor, *Proceedings of the 3rd International Conference on Artificial Intelligence Planning Systems (AIPS-96)*, pp. 27–34. AAAI Press.

Blythe, J. (1999). Decision-theoretic planning. *AI Magazine* 20.

Bonasso, P., J. Firby, E. Gat, D. Kortenkamp, D. Miller, and M. Slack (1997). Experiences with an architecture for intelligent, reactive agents. *Journal of Experimental and Theoretical Artificial Intelligence* 9(1).

Borenstein, J., B. Everett, and L. Feng (1996). *Navigating Mobile Robots: Systems and Techniques.* A. K. Peters, Ltd., Wellesley, MA.

Boutilier, C., T. Dean, and S. Hanks (1998). Decision theoretic planning: Structural assumptions and computational leverage. *Journal of AI research* .

Boutilier, C., R. Reiter, M. Soutchanski, and S. Thrun (2000). Decision-theoretic, high-level robot programming in the situation calculus. In *Proceedings of the AAAI National Conference on Artificial Intelligence*, Austin, TX.

Bradtke, Steven J. and Michael O. Duff (1995). Reinforcement learning methods for continuous-time Markov decision problems. In Tesauro, G., D. Touretzky, and T. Leen, editors, *Advances in Neural Information Processing Systems*, Vol. 7, pp. 393–400. The MIT Press.

Bratman, M. (1987). *Intention, Plans, and Practical Reason.* Harvard University Press, Cambridge, Massachusetts.

Bratman, M., D. Israel, and M. Pollack (1988). Plan and resource-bounded practical reasoning. *Computational Intelligence* 4: 349–355.

Brooks, R. (1986). A robust layered control system for a mobile robot. *IEEE Journal of Robotics and Automation* pp. 14–23.

Burgard, W., A.B. Cremers, D. Fox, D. Hähnel, G. Lakemeyer, D. Schulz, W. Steiner, and S. Thrun (1998). The interactive museum tour-guide robot. In *Proceedings of the Fifteenth National Conference on Artificial Intelligence (AAAI'98).*

Burgard, W., A.B. Cremers, D. Fox, D. Hähnel, G. Lakemeyer, D. Schulz, W. Steiner, and S. Thrun (2000). Experiences with an interactive museum tour-guide robot. *Artificial Intelligence* 114(1-2).

Burgard, W., D. Fox, and S. Thrun (1997). Active mobile robot localization. In *Proceedings of the Fifteenth International Joint Conference on Artificial Intelligence*, Nagoya, Japan.

Burgard, Wolfram, Dieter Fox, and Daniel Hennig (1997). Fast grid-based position tracking for mobile robots. In *Proceedings of the 21th German Conference on Artificial Intelligence (KI 97), Freiburg, Germany.* Springer Verlag.

Bylander, T. (1991). Complexity results for planning. In *Proceedings of the Twelfth International Joint Conference on Artificial Intelligence*, pp. 274–279, Sidney, Australia.

Canny, J. (1986). A computational approach to edge detection. *IEEE Transactions on Pattern Analysis and Machine Intelligence (PAMI)* 8(6): 679–698.

Cohen, P. and H. Levesque (1995). Communicative actions for artificial agents. In *Proceedings of the International Conference on Multi-Agent Systems*, Cambridge, Ma. AAAI Press.

Cohen, P. and C. Perrault (1979). Elements of a plan-based theory of speech acts. *Cognitive Science* 3(3): 177–212.

Cole, R., J. Mariani, H. Uszkoreit, A. Zaenen, and V. Zue (1997). *Survey of the State of the Art in Human Language Technology.* Cambridge University Press and Giardini. `http://www.cse.ogi.edu/CSLU/HLTsurvey/HLTsurvey.html`.

Crowley, J. and H. Christensen (1995). *Vision as Process.* Springer.

Dean, T. and M. Boddy (1988). An analysis of time-dependent planning. In *Proceedings of the Seventh National Conference on Artificial Intelligence*, pp. 49–54, St. Paul, MN.

Dean, T., J. Firby, and D. Miller (1988). Hierarchical planning involving deadlines, travel time and resources. *Computational Intelligence* 4(4): 381–398.

Dean, T. and M. Wellmann (1991). *Planning and Control.* Morgan Kaufmann Publishers, San Mateo, CA.

Despouys, O. and F. Ingrand (1999). Propice-plan: Toward a unified framework for planning and execution. In *Recent Advances in AI Planning. Proceedings of the 1999 European Conference on Planning.* Springer Publishers.

Drabble, B. (1993). Excalibur: a program for planning and reasoning with processes. *Artificial Intelligence* 62: 1–40.

Draper, D., S. Hanks, and D. Weld (1994). Probabilistic planning with information gathering and contingent execution. In Hammond, K., editor, *Proc. 2nd. Int. Conf. on AI Planning Systems.* Morgan Kaufmann.

Finin, Tim, Yannis Labrou, and James Mayfield (1995). *Software Agents*, chapter KQML as an agent communication language. MIT Press, Cambridge, Ma.

Firby, J. (1987). An investigation into reactive planning in complex domains. In *Proceedings of the Sixth National Conference on Artificial Intelligence*, pp. 202–206, Seattle, WA.

Firby, J. (1989). Adaptive Execution in Complex Dynamic Worlds. Technical report 672, Yale University, Department of Computer Science.

Firby, J. (1992). Building symbolic primitives with continuous control routines. In Hendler, J., editor, *Proceedings of the First International Conference on AI Planning Systems*, pp. 62–69, Morgan Kaufmann.

Firby, J. (1994). Task networks for controlling continuous processes. In Hammond, Kris, editor, *Proceedings of the Second International Conference on AI Planning Systems*, pp. 49–54, Morgan Kaufmann.

Firby, J., R. Kahn, P. Prokopowitz, and M. Swain (1995). An architecture for vision and action. In Mellish, C., editor, *Proc. of the $14^{th}$ IJCAI*, pp. 72–79.

Firby, J., P. Prokopowicz, M. Swain, R. Kahn, and F. Franklin (1996). Programming CHIP for the IJCAI-95 robot competition. *AI Magazine* .

Fleury, S., M. Herrb, and R. Chatila (1994). Design of a modular architecture for autonomous robot. In Straub, E. and R. Sipple, editors, *Proceedings of the International Conference on Robotics and Automation. Volume 4*, pp. 3508–3514, Los Alamitos, CA, USA. IEEE Computer Society Press.

Floreano, D. and F. Mondada (1998). Evolutionary neurocontrollers for autonomous mobile robots. *Neural Networks* 11: 1461–178.

Forbus, K. (1984). Qualitative process theory. *Artificial Intelligence* 24: 85–168.

Fox, D., W. Burgard, F. Dellaert, and S. Thrun (1999). Monte Carlo localization: Efficient position estimation for mobile robots. In *Proceedings of the Sixteenth National Conference on Artificial Intelligence*, Orlando, FL.

Fox, D., W. Burgard, and S. Thrun (1997). The dynamic window approach to collision avoidance. *IEEE Robotics and Automation Magazine* .

Gapp, K.-P. (1994). Basic meanings of spatial relations: Computation and evaluation in 3d space. In *Proceedings of the Twelfth National Conference on Artificial Intelligence*, pp. 1393–1398, Seattle, WA.

Georgeff, M. and F. Ingrand (1989). Decision making in an embedded reasing system. In *Proceedings of the Eleventh International Joint Conference on Artificial Intelligence*, pp. 972–978, Detroit, MI.

Georgeff, M. and F. Ingrand (1990). Real-time reasoning: The monitoring and control of spacecraft systems. AI center technical note SRI AI 478, SRI International.

Ghallab, M. and H. Laruelle (1994). Representation and control in IxTeT, a temporal planner. In Hammond, Kris, editor, *Second International Conference on AI Planning Systems*, pp. 61–67, Morgan Kaufmann.

Giacomo, G. De, Y. Lesperance, and H. Levesque (1997). Reasoning about concurrent execution, prioritized interrupts, and exogene ous actions in the situation calculus. In *Proceedings of the Fifteenth International Joint Conference on Artificial Intelligence*, Nagoya, Japan.

Goel, A., E. Stroulia, Z. Chen, and P. Rowland (1997). Model-based reconfiguration of schema-based reactive control architectures. In *Proceedings of the AAAI Fall Symposium on Model-Based Autonomy.*

Goodwin, R. and R. Simmons (1992). Rational handling of multiple goals for mobile robots. In Hendler, J., editor, *Artificial Intelligence Planning Systems: Proceedings of the First International Conference (AIPS 92)*, pp. 70–77, College Park, Maryland, USA. Morgan Kaufmann.

Grosskreutz, H. and G. Lakemeyer (2000a). cc-golog: Towards more realistic logic-based robot controllers. In *Proceedings of the Seventeenth National Conference on Artificial Intelligence.*

Grosskreutz, H. and G. Lakemeyer (2000b). Turning high-level plans into robot programs in uncertain domains. In *ECAI 2000.*

Haddawy, P. and S. Hanks (1992). Representations for decision-theoretic planning: Utility functions for deadline goals. In Nebel, B., C. Rich, and W. Swartout, editors, *Proceedings of the Third International Conference on Principles of Knowledge Representation and Reasoning*, pp. 71–82, Cambridge, MA. Morgan Kaufmann.

Haddawy, P. and L. Rendell (1990). Planning and decision theory. *The Knowledge Engineering Review* 5: 15–33.

Hähnel, D., W. Burgard, and G. Lakemeyer (1998). Golex - bridging the gap between logic (GOLOG) and a real robot. In *Proceedings of the 22nd German Conference on Artificial Intelligence (KI'98).*

Haigh, K. and M. Veloso (1997). High-level planning and low-level execution: Towards a complete robotic agent. In Johnson, L., editor, *Proceedings of the First International Conference on Autonomous Agents*, pp. 363–370, Marina del Rey, CA. (New York, NY: ACM Press).

Haigh, K. and M. Veloso (1998). Learning situation-dependent costs: Improving planning from probabilistic robot execution. In *Second Int. Conf. on Autonomous Agents 98.*

Hanks, S. (1990). Practical temporal projection. In *Proc. of AAAI-90*, pp. 158–163.

Hanks, S., D. Madigan, and J. Gavrin (1995). Probabilistic temporal reasoning with endogenous change. In *Proceedings of the $11^{th}$ Conference on Uncertainty in Artificial Intelligence.*

Hayes, P. (1985). The second naive physics manifesto. In Hobbs, J. R. and R. C. Moore, editors, *Formal Theories of the Commonsense World*, pp. 1–36. Ablex, Norwood, NJ.

Hendrix, G. (1973). Modeling simultaneous actions and continuous processes. *Artificial Intelligence* 4: 145–180.

Horswill, I. (1995). Analysis of adaptation and environment. *Artificial Intelligence* 73: 1–30.

Horswill, I. (1996). Integrated systems and naturalistic tasks. In: Strategic Directions in Computing Research AI Working Group.

Horvitz, E. (1988). Reasoning under varying and uncertain resource constraints. In *Proceedings of the Seventh National Conference on Artificial Intelligence*, pp. 111–116, St. Paul, MN.

Howe, A. and P. Cohen (1992). Isolating dependencies on failure by analyzing execution traces. In Hendler, J., editor, *AIPS-92: Proc. of the First International Conference on Artificial Intelligence Planning Systems*, pp. 277–278. Kaufmann, San Mateo, CA.

Ingrand, F., R. Chatila, R. Alami, and F. Robert (1996). PRS: A high level supervision and control language for autonomous mobile robots. In *Proc. of the IEEE Int. Conf. on Robotics and Automation*, pp. 43–49, Minneapolis.

Ingrand, F., M. Georgeff, and A. Rao (1992). An architecture for real-time reasoning and system control. *IEEE Expert* 7(6).

Kabanza, F., M. Barbeau, and R. St-Denis (1997). Planning control rules for reactive agents. *Artificial Intelligence* 95: 67–113.

Kaelbling, L., A. Cassandra, and J. Kurien (1996). Acting under uncertainty: Discrete bayesian models for mobile-robot navigation. In *Proceedings of the IEEE/RSJ International Conference on Intelligent Robots and Systems.*

Koenig, S. and R. Simmons (1994). How to make reactive planners risk-sensitive. In *Proceedings of the Second International Conference on AI Planning Systems*, pp. 293–304.

Konolige, K., K. Myers, E. Ruspini, and A. Saffiotti (1997). The Saphira architecture: A design for autonomy. *Journal of Experimental and Theoretical Artificial Intelligence* 9(2).

Kortenkamp, D. and T. Weymouth (1994). Topological mapping for mobile robots using a combination of sonar and vision sensing. In *Proceedings of the Twelfth National Conference on Artificial Intelligence*, pp. 979–984, Seattle, WA.

Koza, J. (1992). *Genetic Programming.* MIT Press, Cambridge, MA.

Kushmerick, N., S. Hanks, and D. Weld (1995). An algorithm for probabilistic planning. *Artificial Intelligence* 76: 239–286.

Laborie, P. and M. Ghallab (1995). Planning with sharable resource constraints. In *Proc. of the 14th International Joint Conference on Artificial Intelligence (IJCAI-95)*, pp. 28–34. Morgan Kaufmann.

Laird, J., P. Rosenbloom, and A. Newell (1986). Chunking in soar: the anatomy of a general learning mechanism. *Machine Learning* 1: 11–46.

Lakemeyer, G. (1999). On sensing and off-line interpreting in golog. In Levesque, H. and F. Pirri, editors, *Logical Foundations for Cognitive Agents.* Springer.

Latombe, J.-C. (1991). *Robot Motion Planning.* Kluwer Academic Publishers, Boston, MA.

Leash, N., N. Martin, and J. Allen (1998). Improving big plans. In *Proceedings of the Fifteenth National Conference on Artificial Intelligence (AAAI-98).*

Levesque, H., R. Reiter, Y. Lespérance, F. Lin, and R. Scherl (1997). Golog: A logic programming language for dynamic domains. *Journal of Logic Programming* 31: 59–84.

Levesque, H. J. (1996). What is planning in the presence of sensing. In *Proceedings of the Thirteenth National Conference on Artificial Intelligence*, Portland, OR.

Lochbaum, K. E., B. J. Grosz, and C. L. Sidner (1990). Models of plans to support communication: An initial report. In *Proc. of AAAI-90*, pp. 485–490, Boston, MA.

Lyons, D. and A. Hendriks (1992). A practical approach to integrating reaction and deliberation. In Hendler, J., editor, *Proceedings of the First International Conference on AI Planning Systems*, pp. 153–162, Washington, DC.

Mahadevan, S. (1996). Machine learning for robots: A comparison of different paradigms. In *Workshop on Towards Real Autonomy , IEEE/RSJ International Conference on Intelligent Robots and Systems (IROS-96)*, Osaka, Japan.

McAllester, D. and D. Rosenblitt (1991). Systematic nonlinear planning. In *Proceedings of the Ninth National Conference on Artificial Intelligence*, pp. 634–639, Anaheim, CA.

McCarthy, J. (1963). Situations, actions, and causal laws. Technical report, Stanford University. Reprinted 1968 in Semantic Information Processing (M.Minske ed.), MIT Press.

McDermott, D. (1977). Flexibility and efficiency in a computer program for designing circuits. Technical report AI-TR-402, AI Lab, MIT.

McDermott, D. (1978). Planning and acting. *Cognitive Science* 2(2): 78–109.

McDermott, D. (1985a). The duck manual. Research Report YALEU/DCS/RR-399, Yale University.

McDermott, D. (1985b). Reasoning about plans. In Hobbs, J.R. and R.C. Moore, editors, *Formal Theories of the Commonsense World*, pp. 269–317. Ablex, Norwood, NJ.

McDermott, D. (1991). A Reactive Plan Language. Research Report YALEU/DCS/RR-864, Yale University.

McDermott, D. (1992a). Robot planning. *AI Magazine* 13(2): 55–79.

McDermott, D. (1992b). Transformational planning of reactive behavior. Research Report YALEU/DCS/RR-941, Yale University.

McDermott, D. (1997). An algorithm for probabilistic, totally-ordered temporal projection. In Stock, O., editor, *Spatial and Temporal Reasoning.* Kluwer Academic Publishers, Dordrecht.

McVey, C., E. Atkins, E. Durfee, and K. Shin (1997). Development of iterative real-time scheduler to planner feedback. In *"Proceedings of the 15th Int. Joint Conf. on Artificial Intelligence (IJCAI'87)"*, pp. 1267–1272.

Miller, G., E. Galanter, and K. Pribram (1960). *Plans and the structure of behaviour.* Holt, Rinehart and Winston, New York.

Mitchell, T. (1990). Becoming increasingly reactive (mobile robots). In *Proceedings of the Eighth National Conference on Artificial Intelligence*, pp. 1051–1058, Boston, MA. MIT Press.

Mitchell, T. and S. Thrun (1993). Explanation-based neural network learning for robot control. In Giles, C. L., S. J. Hanson, and J. D. Cowan, editors, *Advances in Neural Information Processing Systems 5, Proceedings of the IEEE Conference in Denver*, San Mateo, CA. Morgan Kaufmann.

Moore, A. (1994). The parti-game algorithm for variable resolution reinforcement learning in multidimensional state-spaces. In Cowan, Jack D., Gerald Tesauro, and Joshua Alspector, editors, *Advances in Neural Information Processing Systems*, Vol. 6, pp. 711–718. Morgan Kaufmann Publishers, Inc.

Muscettola, N., P. Morris, B. Pell, and B. Smith (1998a). Issues in temporal reasoning for autonomous control systems. In Sycara, K. and M. Wooldridge, editors, *Proceedings of the 2nd International Conference on Autonomous Agents (AGENTS'98)*, pp. 362–368, New York. ACM Press.

Muscettola, N., P. Nayak, B. Pell, and B. Williams (1998b). Remote Agent: to go boldly where no AI system has gone before. *Artificial Intelligence* 103(1–2): 5–47.

Myers, K. (1996). A procedural knowledge approach to task-level control. In Drabble, B., editor, *Proceedings of the Third International Conference on AI Planning Systems*, pp. 158–165, Edinburgh, GB. AAAI Press.

Nebel, B. and J. Koehler (1995). Plan reuse versus plan generation: a theoretical and empirical analysis. *Artificial Intelligence* 76(1–2): 427–454.

Nilsson, Nils J. (1984). Shakey the robot. Technical Note 323, SRI International, Menlo Park, California.

Norvig, P. (1992). *Paradigms of Artificial Intelligence Programming: Case Studies in Common Lisp.* Morgan Kaufmann, San Mateo, CA.

Oates, T., M. Schmill, and P. Cohen (2000). Identifying qualitatively different outcomes of actions: Gaining autonomy through learning. In *Proceedings of the Fourth International Conference on Autonomous Agents*, pp. 110–111, Barcelona, Spain. ACM Press.

Parr, R. (1998). Flexible decomposition algorithms for weakly coupled Markov decision problems. In Cooper, Gregory F. and Serafín Moral, editors, *Proceedings of the 14th Conference on Uncertainty in Artificial Intelligence (UAI-98)*, pp. 422–430, San Francisco. Morgan Kaufmann.

Parr, R. and S. Russell (1998). Reinforcement learning with hierarchies of machines. In Jordan, Michael I., Michael J. Kearns, and Sara A. Solla, editors, *Advances in Neural Information Processing Systems*, Vol. 10. The MIT Press.

Passino, K. and P. Antsaklis (1989). A system and control-theoretic perspective on artificial intelligence planning systems. *Applied Artificial Intelligence* 3: 1–32.

Pell, B., D. Bernard, S. Chien, E. Gat, N. Muscettola, P. Nayak, M. Wagner, and B. Williams (1997). An autonomous spacecraft agent prototype. In *Proceedings of the First International Conference on Autonomous Agents.*

Pell, B., E. Gamble, E. Gat, R. Keesing, J. Kurien, W. Millar, P. Nayak, C. Plaunt, and B. Williams (1998). A hybrid procedural/deductive executive for autonomous spacecraft. In Sycara, K. and M. Wooldridge, editors, *Proceedings of the 2nd International Conference on Autonomous Agents (AGENTS-98)*, pp. 369–376, New York. ACM Press.

Pell, B., E. Gat, R. Keesing, N. Muscettola, and B. Smith (1997a). Plan execution for autonomous spacecraft. In Pollack, M., editor, *Proceedings of the 15th International Joint Conference on Artificial Intelligence (IJCAI-97)*, pp. 1234–1239. Morgan Kaufmann.

Pell, B., E. Gat, R. Keesing, N. Muscettola, and B. Smith (1997b). Robust periodic planning and execution for autonomous spacecraft. In *Proceedings of the 15th International Joint Conference on Artificial Intelligence (IJCAI-97)*, pp. 1234–1239, San Francisco. Morgan Kaufmann Publishers.

Pereira, F. and D. Warren (1980). Definite clause grammars for language analysis. In *Artificial Intelligence*, pp. 231–278.

Pinto, J. (1994). Temporal Reasoning in the Situation Calculus. Ph.D. diss., Department of Computer Science, University of Toronto, Toronto, Ontario, Canada.

Pollack, M. and J. Horty (1999). There's more to life than making plans: Plan management in dynamic, multi-agent environments. *AI Magazine* 20(4): 71–84.

Pollack, M. and M. Ringuette (1990). Introducing the tileworld: Experimentally evaluating agent architectures. In *Proceedings of the Eighth National Conference on Artificial Intelligence*, pp. 183–189, Boston, MA.

Prokopowicz, P., M. Swain, J. Firby, and R. Kahn (1996). GARGOYLE: An environment for real-time, context-sensitive active vision. In *Proc. of the Fourteenth National Conference on Artificial Intelligence*, pp. 930–937.

Ramsey, F. (1931). Truth and probability. In Braithwaite, R. B., editor, *The Foundations of Mathematics: Collected Papers of Frank P. Ramsey*, pp. 156–198. Routledge and Kegan Paul, London.

Rao, A. and M. Georgeff (1992). An abstract architecture for rational agents. In Nebel, B., C. Rich, and W. Swartout, editors, *Principles of Knowledge Representation and Reasoning: Proc. of the Third International Conference (KR'92)*, pp. 439–449, San Mateo, CA. Kaufmann.

Rasure and Young (1992). An open environment for image processing software development. In *Proceedings of the SPIE/IS&T Symposium on Electronic Imaging (SPIE-92)*.

Reichman, R. (1981). Plain-speaking: A theory and grammar of spontaneous discourse. Ph.D. diss., Harvard University, Cambridge, Massachusetts.

Reiter, R. (1996). Natural actions, concurrency and continuous time in the situation calculus. In *Proceedings of the 5th International Conference on Principles of Knowledge Representation and Reasoning (KR-96)*, pp. 2–13.

Rich, E. and K. Knight (1991). *Artificial Intelligence.* McGraw Hill, New York.

Rohanimanesh, K. and S. Mahadevan (2001). Decision-theoretic planning with concurrent temporally extended actions. In *17th Conference on Uncertainty in Artificial Intelligence (UAI)*, University of Washington, Seattle, WA.

Russell, S. and P. Norvig (1995). *Artificial Intelligence: A Modern Approach.* Prentice-Hall, Englewood Cliffs, NJ.

Russell, S. and D. Subramanian (1995). Provably bounded-optimal agents. *Journal of Artificial Intelligence Research* 3.

Russell, S. and E. Wefald (1991). *Do the right thing: Studies in limited rationality.* MIT Press, Cambridge, MA.

Saffiotti, A., N. Helft, K. Konolige, J. Lowrance, K. Myers, D. Musto, E. Ruspini, and L. Wesley (1993). A fuzzy controller for Flakey, the robot. In *Proceedings of the Eleventh National Conference on Artificial Intelligence*, pp. 864–864, Cambridge, MA. AAAI Press.

Santamaria, J. and A. Ram (1997). Learning of parameter-adaptive reactive controllers for robotic navigation. In *Proceedings of the World Multiconference on Systemics, Cybernetics, and Informatics*, Caracas, Venezuela.

Schoppers, M. (1987). Universal plans for reactive robots in unpredictable environments. In *Proceedings of the Tenth International Joint Conference on Artificial Intelligence*, pp. 1039–1046, Milan, Italy. Morgan Kaufmann.

Schulz, D. and W. Burgard (2001). Probabilistic state estimation of dynamic objects with a moving mobile robot. *Robotics and Autonomous Systems* 34(2-3): 107–115.

Searle, John R. (1969). *Speech Acts: An Essay in the Philosophy of Language.* Cambridge University Press, Cambridge.

Simmons, R. (1992). The role of associational and causal reasoning in problem solving. *AI Journal* 53.

Simmons, R. (1994). A robust layered control system for a mobile robot. *IEEE Journal of Robotics and Automation* pp. 34–43.

Simmons, R., R. Goodwin, K. Haigh, S. Koenig, and J. O'Sullivan (1997a). A modular architecture for office delivery robots. In *Proceedings of the First International Conference on Autonomous Agents*, pp. 245–252.

Simmons, R., R. Goodwin, K. Haigh, S. Koenig, J. O'Sullivan, and M. Veloso (1997b). Xavier: Experience with a layered robot architecture. *ACM magazine* Intelligence .

Simmons, R. and S. Koenig (1995). Probabilistic robot navigation in partially observable environments. In *Proceedings of the International Joint Conference on Artificial Intelligence*, pp. 1080–1087.

Simon, H. A. (1982). Models of bounded rationality. In *Vol II: Behavioural economics and business organization.* MIT press.

Smith, D., J. Frank, and A. Jonsson (2000). Bridging the gap between planning and scheduling. *Knowledge Engineering Review* 15(1).

Steels, L. (1984). Design requirements for knowledge representation systems. In Laubsch, J., editor, *Proceedings of GWAI-84*, pp. 1–19, Wingst, Germany.

Stopp, E., K. Gapp, G. Herzog, T. Längle, and T. Lüth (1994). Utilizing spatial relations for natural language access to an autonomous mobile robot. In Nebel, B. and L. Dreschler-Fischer, editors, *KI-94: Advances in Artificial Intelligence.*, pp. 39–50, Berlin, Heidelberg. Springer.

Suchman, L. (1985). *Plans and Situated Actions.* Cambridge University Press.

Sussman, G. (1977). *A Computer Model of Skill Acquisition*, Vol. 1 of *Aritficial Intelligence Series.* American Elsevier, New York, NY.

Sussman, G. (1990). The virtuous nature of bugs. In Allen, J., J. Hendler, and A. Tate, editors, *Readings in Planning*, pp. 111–117. Kaufmann, San Mateo, CA.

Sutton, R. and A. Barto (1998). *Reinforcement Learning: an Introduction.* MIT Press.

Sutton, R., D. Precup, and S. Singh (1998). Between mdps and semi-mdps: Learning, planning, and representing knowledge at multiple temporal scales. *Journal of AI Research* .

Sutton, R., D. Precup, and S. Singh (1999). Between mdps and semi-mdps: A framework for temporal abstraction in reinforcement learning. *Artificial Intelligence 112(1-2)* pp. 181–211.

Thrun, S. (1999). Monte Carlo POMDPs. In *Proceedings of Conference on Neural Information Processing Systems (NIPS).*

Thrun, S., M. Beetz, M. Bennewitz, A.B. Cremers, F. Dellaert, D. Fox, D. Hähnel, C. Rosenberg, N. Roy, J. Schulte, and D. Schulz (2000). Probabilistic algorithms and the interactive museum tour-guide robot Minerva. *International Journal of Robotics Research.*

Thrun, S., M. Bennewitz, W. Burgard, A.B. Cremers, F. Dellaert, D. Fox, D. Haehnel, C. Rosenberg, N. Roy, J. Schulte, and D. Schulz (1999). Minerva: A second generation mobile tour-guide robot. In *Proceedings of the IEEE International Conference on Robotics and Automation (ICRA'99).*

Thrun, S., A. Bücken, W. Burgard, D. Fox, T. Fröhlinghaus, D. Hennig, T. Hofmann, M. Krell, and T. Schmidt (1998). Map learning and high-speed navigation in RHINO. In Kortenkamp, D., R.P. Bonasso, and R. Murphy, editors, *AI-based Mobile Robots: Case studies of successful robot systems.* MIT Press, Cambridge, MA.

Torrance, Mark C. and Lynn Andrea Stein (1997). Communicating with martians (and robots). Technical report, MIT Artificial Intelligence Laboratory.

Towell, G. and J. Shavlik (1989). An approach to combining explanation-based and neural learning algorithms. *Connection Science* 1: 233–255.

Ullman, S. (1984). Visual routines. *Cognition* pp. 97–159.

Weis, T. (1997). Resource-adaptive action planning in a dialogue system for repair support. In Nebel, B., editor, *Proceedings der 21. Deutschen Jahrestagung für Künstliche Intelligenz*, Berlin, New Yor. Springer.

Wilensky, R. (1983). *Planning and Understanding. A Computational Approach to Human Reasoning.* Addison-Wesley Publishing Company.

Wilkins, D. and K. Myers (1995). A common knowledge representation for plan generation and reactive execution. *Journal of Logic and Computation* 5(6): 731–761.

Wilkins, D., K. Myers, J. Lowrance, and L. Wesley (1995). Planning and reacting in uncertain and dynamic environments. *Journal of Experimental and Theoretical AI* 7(1): 197–227.

Williams, B. and P. Nayak (1996). A model-based approach to reactive self-configuring systems. In *Proceedings of the Thirteenth National Conference on Artificial Intelligence and the Eighth Innovative Applications of Artificial Intelligence Conference*, pp. 971–978, Menlo Park. AAAI Press / MIT Press.

Williams, B. and P. Nayak (1997). A reactive planner for a model-based executive. In *Proceedings of the 15th International Joint Conference on Artificial Intelligence (IJCAI-97)*, pp. 1178–1185, San Francisco. Morgan Kaufmann Publishers.

Williamson, M. and S. Hanks (1994). Utility-directed planning. In *Proceedings of the Twelfth National Conference on Artificial Intelligence*, p. 1498, Seattle, WA.

Winograd, Terry (1972). Understanding natural language. *Cognitive Psychology* 3(1). Reprinted as a book by Academic Press.

Zilberstein, S. (1995). On the utility of planning. In Pollack, M., editor, *SIGART Bulletin Special Issue on Evaluating Plans, Planners, and Planning Systems*, Vol. 6. ACM.

Zilberstein, S. (1996). Using anytime algorithms in intelligent systems. *AI Magazine* 17(3): 73–83.

Zilberstein, S. and S. Russell (1995). Optimal composition of real-time systems. *Artificial Intelligence* 79(2).

# Lecture Notes in Artificial Intelligence (LNAI)

Vol. 2381: U. Egly, C.G. Fermüller (Eds.), Automated Reasoning with Analytic Tableaux and Related Methods. Proceedings, 2002. X, 341 pages. 2002.

Vol. 2385: J. Calmet, B. Benhamou, O. Caprotti, L. Henocque, V. Sorge (Eds.), Artificial Intelligence, Automated Reasoning, and Symbolic Computation. Proceedings, 2002. XI, 343 pages. 2002.

Vol. 2389: E. Ranchhod, N.J. Mamede (Eds.), Advances in Natural Language Processing. Proceedings, 2002. XII, 275 pages. 2002.

Vol. 2392: A. Voronkov (Ed.), Automated Deduction - CADE-18. Proceedings, 2002. XII, 534 pages. 2002.

Vol. 2393: U. Priss, D. Corbett, G. Angelova (Eds.), Conceptual Structures: Integration and Interfaces. Proceedings, 2002. XI, 397 pages. 2002.

Vol. 2394: P. Perner (Ed.), Advances in Data Mining. VII, 109 pages. 2002.

Vol. 2403: Mark d'Inverno, M. Luck, M. Fisher, C. Preist (Eds.), Foundations and Applications of Multi-Agent Systems. Proceedings, 1996-2000. X, 261 pages. 2002.

Vol. 2407: A.C. Kakas, F. Sadri (Eds.), Computational Logic: Logic Programming and Beyond. Part I. XII, 678 pages. 2002.

Vol. 2408: A.C. Kakas, F. Sadri (Eds.), Computational Logic: Logic Programming and Beyond. Part II. XII, 628 pages. 2002.

Vol. 2413: K. Kuwabara, J. Lee (Eds.), Intelligent Agents and Multi-Agent Systems. Proceedings, 2002. X, 221 pages. 2002.

Vol. 2416: S. Craw, A. Preece (Eds.), Advances in Case-Based Reasoning. Proceedings, 2002. XII, 656 pages. 2002.

Vol. 2417: M. Ishizuka, A. Sattar (Eds.), PRICAI 2002: Trends in Artificial Intelligence. Proceedings, 2002. XX, 623 pages. 2002.

Vol. 2424: S. Flesca, G. Ianni (Eds.), Logics in Artificial Intelligence. Proceedings, 2002. XIII, 572 pages. 2002.

Vol. 2427: M. Hannebauer, Autonomous Dynamic Reconfiguration in Multi-Agent Systems. XXI, 284 pages. 2002.

Vol. 2430: T. Elomaa, H. Mannila, H. Toivonen (Eds.), Machine Learning: ECML 2002. Proceedings, 2002. XIII, 532 pages. 2002.

Vol. 2431: T. Elomaa, H. Mannila, H. Toivonen (Eds.), Principles of Data Mining and Knowledge Discovery. Proceedings, 2002. XIV, 514 pages. 2002.

Vol. 2443: D. Scott (Ed.), Artificial Intelligence: Methodology, Systems, and Applications. Proceedings, 2002. X, 279 pages. 2002.

Vol. 2445: C. Anagnostopoulou, M. Ferrand, A. Smaill (Eds.), Music and Artificial Intelligence. Proceedings, 2002. VIII, 207 pages. 2002.

Vol. 2446: M. Klusch, S. Ossowski, O. Shehory (Eds.), Cooperative Information Agents VI. Proceedings, 2002. XI, 321 pages. 2002.

Vol. 2447: D.J. Hand, N.M. Adams, R.J. Bolton (Eds.), Pattern Detection and Discovery. Proceedings, 2002. XII, 227 pages. 2002.

Vol. 2448: P. Sojka, I. Kopecÿek, K. Pala (Eds.), Text, Speech and Dialogue. Proceedings, 2002. XII, 481 pages. 2002.

Vol. 2464: M. O'Neill, R.F.E. Sutcliffe, C. Ryan, M. Eaton, N. Griffith (Eds.), Artificial Intelligence and Cognitive Science. Proceedings, 2002. XI, 247 pages. 2002.

Vol. 2466: M. Beetz, J. Hertzberg, M. Ghallab, M.E. Pollack (Eds.), Advances in Plan-Based Control of Robotic Agents. Proceedings, 2001. VIII, 291 pages. 2002.

Vol. 2473: A. Gómez-Pérez, V.R. Benjamins, Knowledge Engineering and Knowledge Management. Proceedings, 2002. XI, 402 pages. 2002.

Vol. 2475: J.J. Alpigini, J.F. Peters, A. Skowron, N. Zhong (Eds.), Rough Sets and Current Trends in Computing. Proceedings, 2002. XV, 640 pages. 2002.

Vol. 2479: M. Jarke, J. Koehler, G. Lakemeyer (Eds.), KI 2002: Advances in Artificial Intelligence. Proceedings, 2002. XIII, 327 pages. 2002.

Vol. 2484: P. Adriaans, H. Fernau, M. van Zaanen (Eds.), Grammatical Inference: Algorithms and Applications. Proceedings, 2002. IX, 315 pages. 2002.

Vol. 2499: S.D. Richardson (Ed.), Machine Translation: From Research to Real Users. Proceedings, 2002. XXI, 254 pages. 2002.

Vol. 2504: M.T. Escrig, F. Toledo, E. Golobardes (Eds.), Topics in Artificial Intelligence. Proceedings 2002. XI, 432 pages. 2002.

Vol. 2507: G. Bittencourt, G.L. Ramalho (Eds.), Advances in Artificial Intelligence. Proceedings, 2002. XIII, 418 pages. 2002.

Vol. 2514: M. Baaz, A. Voronkov (Eds.), Logic for Programming, Artificial Intelligence, and Reasoning. Proceedings 2002. XIII, 465 pages. 2002.

Vol. 2522: T. Andreasen, A. Motro, H. Christiansen, H. Legind Larsen (Eds.), Flexible Query Answering. Proceedings 2002. XI, 386 pages. 2002.

Vol. 2527: F.J. Garijo, J.C. Riquelme, M. Toro (Eds.), Advances in Artificial Intelligence – IBERAMIA 2002. Proceedings 2002. XVIII, 955 pages. 2002.

Vol. 2533: N. Cesa-Bianchi, M. Numao, R. Reischuk (Eds.), Algorithmic Learning Theory. Proceedings 2002. XI, 415 pages. 2002.

Vol. 2541: T. Barkowsky, Mental Representation and Processing of Geographic Knowledge. X, 174 pages. 2002.

Vol. 2554: M. Beetz, Plan-Based Control of Robotic Agents. XI, 191 pages. 2002.

Vol. 2557: B. McKay, J. Slaney (Eds.), AI 2002: Advances in Artificial Intelligence. Proceedings 2002. XV, 730 pages. 2002.

Vol. 2569: D. Gollmann, G. Karjoth, M. Waidner (Eds.), Computer Security – ESORICS 2002. Proceedings, 2002. XIII, 648 pages. 2002.

# Lecture Notes in Computer Science

Vol. 2513: R. Deng, S. Qing, F. Bao, J. Zhou (Eds.), Information and Communications Security. Proceedings, 2002. XII, 496 pages. 2002.

Vol. 2514: M. Baaz, A. Voronkov (Eds.), Logic for Programming, Artificial Intelligence, and Reasoning. Proceedings, 2002. XIII, 465 pages. 2002. (Subseries LNAI).

Vol. 2515: F. Boavida, E. Monteiro, J. Orvalho (Eds.), Protocols and Systems for Interactive Distributed Multimedia. Proceedings, 2002. XIV, 372 pages. 2002.

Vol. 2516: A. Wespi, G. Vigna, L. Deri (Eds.), Recent Advances in Intrusion Detection. Proceedings, 2002. X, 327 pages. 2002.

Vol. 2517: M.D. Aagaard, J.W. O'Leary (Eds.), Formal Methods in Computer-Aided Design. Proceedings, 2002. XI, 399 pages. 2002.

Vol. 2518: P. Bose, P. Morin (Eds.), Algorithms and Computation. Proceedings, 2002. XIII, 656 pages. 2002.

Vol. 2519: R. Meersman, Z. Tari, et al. (Eds.), On the Move to Meaningful Internet Systems 2002: CoopIS, DOA, and ODBASE. Proceedings, 2002. XXIII, 1367 pages. 2002.

Vol. 2521: A. Karmouch, T. Magedanz, J. Delgado (Eds.), Mobile Agents for Telecommunication Applications. Proceedings, 2002. XII, 317 pages. 2002.

Vol. 2522: T. Andreasen, A. Motro, H. Christiansen, H. Legind Larsen (Eds.), Flexible Query Answering. Proceedings, 2002. XI, 386 pages. 2002. (Subseries LNAI).

Vol. 2525: H.H. Bülthoff, S.-Whan Lee, T.A. Poggio, C. Wallraven (Eds.), Biologically Motivated Computer Vision. Proceedings, 2002. XIV, 662 pages. 2002.

Vol. 2526: A. Colosimo, A. Giuliani, P. Sirabella (Eds.), Medical Data Analysis. Proceedings, 2002. IX, 222 pages. 2002.

Vol. 2527: F.J. Garijo, J.C. Riquelme, M. Toro (Eds.), Advances in Artificial Intelligence - IBERAMIA 2002. Proceedings, 2002. XVIII, 955 pages. 2002. (Subseries LNAI).

Vol. 2528: M.T. Goodrich, S.G. Kobourov (Eds.), Graph Drawing. Proceedings, 2002. XIII, 384 pages. 2002.

Vol. 2529: D.A. Peled, M.Y. Vardi (Eds.), Formal Techniques for Networked and Distributed Sytems - FORTE 2002. Proceedings, 2002. XI, 371 pages. 2002.

Vol. 2532: Y.-C. Chen, L.-W. Chang, C.-T. Hsu (Eds.), Advances in Multimedia Information Processing - PCM 2002. Proceedings, 2002. XXI, 1255 pages. 2002.

Vol. 2533: N. Cesa-Bianchi, M. Numao, R. Reischuk (Eds.), Algorithmic Learning Theory. Proceedings, 2002. XI, 415 pages. 2002. (Subseries LNAI).

Vol. 2534: S. Lange, K. Satoh, C.H. Smith (Ed.), Discovery Science. Proceedings, 2002. XIII, 464 pages. 2002.

Vol. 2535: N. Suri (Ed.), Mobile Agents. Proceedings, 2002. X, 203 pages. 2002.

Vol. 2536: M. Parashar (Ed.), Grid Computing - GRID 2002. Proceedings, 2002. XI, 318 pages. 2002.

Vol. 2537: D.G. Feitelson, L. Rudolph, U. Schwiegelshohn (Eds.), Job Scheduling Strategies for Parallel Processing. Proceedings, 2002. VII, 237 pages. 2002.

Vol. 2538: B. König-Ries, K. Makki, S.A.M. Makki, N. Pissinou, P. Scheuermann (Eds.), Developing an Infrastructure for Mobile and Wireless Systems. Proceedings 2001. X, 183 pages. 2002.

Vol. 2540: W.I. Grosky, F. Plášil (Eds.), SOFSEM 2002: Theory and Practice of Informatics. Proceedings, 2002. X, 289 pages. 2002.

Vol. 2541: T. Barkowsky, Mental Representation and Processing of Geographic Knowledge. X, 174 pages. 2002. (Subseries LNAI).

Vol. 2544: S. Bhalla (Ed.), Databases in Networked Information Systems. Proceedings 2002. X, 285 pages. 2002.

Vol. 2545: P. Forbrig, Q, Limbourg, B. Urban, J. Vanderdonckt (Eds.), Interactive Systems. Proceedings 2002. X, 269 pages. 2002.

Vol. 2546: J. Sterbenz, O. Takada, C. Tschudin, B. Plattner (Eds.), Active Networks. Proceedings, 2002. XIV, 267 pages. 2002.

Vol. 2548: J. Hernández, Ana Moreira (Eds.), Object-Oriented Technology. Proceedings, 2002. VIII, 223 pages. 2002.

Vol. 2549: J. Cortadella, A. Yakovlev, G. Rozenberg (Eds.), Concurrency and Hardware Design. XI, 345 pages. 2002.

Vol. 2550: A. Jean-Marie (Ed.), Advances in Computing Science – ASIAN 2002. Proceedings, 2002. X, 233 pages. 2002.

Vol. 2551: A. Menezes, P. Sarkar (Eds.), Progress in Cryptology – INDOCRYPT 2002. Proceedings, 2002. XI, 437 pages. 2002.

Vol. 2552: S. Sahni, V.K. Prasanna, U. Shukla (Eds.), High Performance Computing – HiPC 2002. Proceedings, 2002. XXI, 735 pages. 2002.

Vol. 2553: B. Andersson, M. Bergholtz, P. Johannesson (Eds.), Natural Language Processing and Information Systems. Proceedings, 2002. X, 241 pages. 2002.

Vol. 2554: M. Beetz, Plan-Based Control of Robotic Agents. XI, 191 pages. 2002. (Subseries LNAI).

Vol. 2555: E.-P. Lim, S. Foo, C. Khoo, H. Chen, E. Fox, S. Urs, T. Costantino (Eds.), Digital Libraries: People, Knowledge, and Technology. Proceedings, 2002. XVII, 535 pages. 2002.

Vol. 2556: M. Agrawal, A. Seth (Eds.), FST TCS 2002: Foundations of Software Technology and Theoretical Computer Science. Proceedings, 2002. XI, 361 pages. 2002.

Vol. 2557: B. McKay, J. Slaney (Eds.), AI 2002: Advances in Artificial Intelligence. Proceedings, 2002. XV, 730 pages. 2002. (Subseries LNAI).

Vol. 2559: M. Oivo, S. Komi-Sirviö (Eds.), Product Focused Software Process Improvement. Proceedings, 2002. XV, 646 pages. 2002.

Vol. 2561: H.C.M. de Swart (Ed.), Relational Methods in Computer Science. Proceedings, 2001. X, 315 pages. 2002.

Vol. 2569: D. Gollmann, G. Karjoth, M. Waidner (Eds.), Computer Security – ESORICS 2002. Proceedings, 2002. XIII, 648 pages. 2002. (Subseries LNAI).